AF779260

Corrosion Resistance of Innovative Composite Coatings Based on Nanoparticles

Corrosion Resistance of Innovative Composite Coatings Based on Nanoparticles

Guest Editors

Marin Kurtela
Ivan Stojanović

MDPI

Basel • Beijing • Wuhan • Barcelona • Belgrade • Novi Sad • Cluj • Manchester

Guest Editors

Marin Kurtela
Faculty of Mechanical Engineering and Naval Architecture
University of Zagreb
Zagreb
Croatia

Ivan Stojanović
Faculty of Mechanical Engineering and Naval Architecture
University of Zagreb
Zagreb
Croatia

Editorial Office

MDPI AG
Grosspeteranlage 5
4052 Basel, Switzerland

This is a reprint of the Special Issue, published open access by the journal *Coatings* (ISSN 2079-6412), freely accessible at: https://www.mdpi.com/journal/coatings/special_issues/VSHY29R2D5.

For citation purposes, cite each article independently as indicated on the article page online and as indicated below:

Lastname, A.A.; Lastname, B.B. Article Title. *Journal Name* **Year**, *Volume Number*, Page Range.

ISBN 978-3-7258-7595-5 (Hbk)
ISBN 978-3-7258-7596-2 (PDF)
https://doi.org/10.3390/books978-3-7258-7596-2

© 2026 by the authors. Articles in this reprint are Open Access and distributed under the Creative Commons Attribution (CC BY) license. The reprint as a whole is distributed by MDPI under the terms and conditions of the Creative Commons Attribution-NonCommercial-NoDerivs (CC BY-NC-ND) license (https://creativecommons.org/licenses/by-nc-nd/4.0/).

Contents

About the Editors

Marin Kurtela

Marin Kurtela was employed from 2011 to 2014 at the Department of Process Management at Brodarski Institut ltd., where as a research assistant and later as a team leader, he participated in the organization and successful implementation of more than 30 different projects, related to the design and performance of turbine control systems. From 2014 to 2016, he was employed in a private company, where as an independent designer of ship equipment and hydraulic systems, he participated in the organization and implementation of several dozen projects related to the design, supervision and construction of deck equipment, both for the domestic shipbuilding industry and for the international market of luxury yachts. Since 2016, he has been working at the Chair of Materials Protection at the Faculty of Mechanical Engineering and Naval Architecture of the University of Zagreb where he earned his PhD in 2021 and currently works as an assistant professor. He has authored and co-authored over 70 scientific and professional papers in the field of corrosion and materials protection, including more than 20 scientific papers published in CC and SCI-expanded journals. In 2020 he became a member of the technical committee TO 504, Corrosion and Protection of Metals and Alloys of the Croatian Standards Institute. His research focuses primarily on corrosion-resistant materials and failure analysis, corrosion inhibitors, and organic coatings.

Ivan Stojanović

Ivan Stojanović earned his PhD from the Faculty of Mechanical Engineering and Naval Architecture at the University of Zagreb, in 2011. He was appointed as Assistant Professor in 2015, then promoted to Associate Professor in 2020, and later to Full Professor in 2025. He currently serves as the Head of the Materials Protection Laboratory. He has authored and co-authored over 120 scientific and professional papers in the field of corrosion and materials protection, including more than 30 scientific papers published in CC and SCI-expanded journals. He also co-authored the university textbook "Corrosion and Coatings", published by the Faculty of Mechanical Engineering and Naval Architecture, University of Zagreb. His research focuses primarily on organic coatings, corrosion inhibitors, corrosion-resistant materials, and failure analysis.

coatings

MDPI

Editorial

Corrosion Resistance of Innovative Composite Coatings Based on Nanoparticles

Marin Kurtela * and Ivan Stojanović

Department of Welded Structures, Faculty of Mechanical Engineering and Naval Architecture, University of Zagreb, 10000 Zagreb, Croatia; ivan.stojanovic@fsb.unizg.hr
* Correspondence: marin.kurtela@fsb.unizg.hr; Tel.: +385-1-6168-343

Organic coatings remain a cornerstone of corrosion protection strategies for metallic substrates due to their cost-effectiveness, ease of processing, and adaptability to diverse industrial applications. These coatings primarily act by forming a protective barrier that impedes the ingress of aggressive species such as chloride ions, moisture, and acidic components, which are notorious for accelerating corrosion damage under service conditions. However, traditional organic coatings often exhibit limited long-term durability in highly aggressive environments, where degradation mechanisms such as blistering, microcracking, increased porosity, and loss of adhesion can compromise performance and accelerate substrate deterioration [1,2]. The global economic impact of corrosion on costs, premature equipment failure, loss of product and plant, highlights the urgent need for improved protective systems that are both highly efficient and environmentally friendly. In this context, nanotechnology has emerged as a powerful enabler for next-generation organic coatings, allowing for the design of multifunctional nanocomposite systems with superior corrosion resistance, mechanical strength, and enhanced service life [3–5]. Nanocomposite coatings, which incorporate finely dispersed nanoparticles into polymer matrices, effectively create tortuous pathways for corrosive agents, reduce defect populations, and improve barrier effectiveness, resulting in significantly enhanced anti-corrosion behaviour compared to conventional coatings [3,6]. Graphene and its derivatives have been extensively studied as reinforcement agents in anticorrosion coatings due to their exceptional impermeability, mechanical stiffness, and large specific surface area, which contribute to enhanced barrier performance and reduced ion transport [5,7]. The incorporation of graphene oxide into epoxy and polyurethane coatings can significantly delay the onset of corrosion and improve coating stability in aggressive chloride environments [8,9]. Complementary research has focused on other nanomaterials, including metal oxides (e.g., ZrO_2, Al_2O_3, and TiO_2), carbon nanotubes (CNTs), and hybrid organic–inorganic clusters, which improve corrosion performance through synergistic effects on coating morphology, adhesion, and electrochemical stability [10–12]. Environmentally sustainable nanocomposite coatings that utilise bio-derived components and green synthesis methods are also gaining attention as efforts increase to minimise the ecological footprint of protective coatings without compromising performance [12,13]. Beyond passive barrier enhancement, smart anticorrosion coatings that incorporate nanocontainers, self-healing agents, and responsive inhibitors capable of releasing corrosion inhibitors upon damage have been developed, thereby enabling active protection mechanisms [4,14]. These smart systems represent a paradigm shift in coating design, as they actively mitigate corrosion progression rather than solely delaying corrosive ingress. Recent advances have also explored multifunctional nanocomposite systems that combine corrosion resistance with antifouling, antibacterial, or thermal-management properties, further expanding the potential applications of advanced coatings in marine, biomedical, and energy sectors [11,15]. Factors such as nanoparticle dispersion, matrix compatibility, coating thickness, and microstructure optimisation are increasingly recognised as key determinants of long-term performance, with numerous studies addressing strategies to control these parameters for enhanced protection [3,7]. Despite considerable progress, substantial challenges remain, particularly in scaling up manufacturing processes, maintaining uniform nanoparticle dispersion, ensuring environmental sustainability, and rigorously validating long-term stability under real-world conditions. Future research should therefore emphasise the development of highly scalable, eco-friendly nanocomposite coatings with integrated smart functionalities, coupled with comprehensive field testing to demonstrate reliable performance in practical applications.

This Special Issue on "Corrosion Resistance of Innovative Composite Coatings Based on Nanoparticles" aimed to bring together cutting-edge research on the synthesis, characterisation, and application of functionalized nanomaterials in advanced anticorrosion coatings. By deepening our understanding of structure–property relationships, corrosion mechanisms, and multifunctional design strategies, we hope to stimulate further innovation and facilitate the translation of laboratory discoveries into commercially viable solutions.

The article by Kurtela et al. investigated the electrochemical corrosion performance of three PACVD coatings—TiN, TiCN, and TiBN—deposited on WC-Co hardmetal substrates. Using a combination of direct current (potentiodynamic and linear polarisation) and alternating current (electrochemical impedance spectroscopy, EIS) techniques in 3.5 wt.% NaCl solution, the coatings' protective abilities were systematically compared. TiN exhibited the highest corrosion resistance, characterised by the most positive open circuit potential, the lowest corrosion current density, and the highest polarisation and charge transfer resistance. TiBN exhibited moderate protection, while TiCN showed comparatively higher corrosion rates and lower electrochemical stability, likely due to its multilayer microstructure facilitating electrolyte penetration. EIS analysis confirmed these trends, showing that TiN's dense, uniform barrier layer effectively impedes charge transfer, whereas TiCN is the most permeable. These results highlight the critical influence of coating architecture, chemical composition, and microstructural properties on corrosion protection, providing insights for optimising multilayer gradient coatings for enhanced durability in aggressive environments [16].

In another article, Aperador et al. investigated the development of silver (Ag)-doped zirconia (ZrO_2) coatings deposited on 316LVM stainless steel via the unbalanced magnetron sputtering technique. The Ar/O_2 ratio during deposition significant influence on the microstructure, tribological behaviour, and corrosion resistance of Ag-ZrO_2 coatings. Increasing oxygen content enhances ZrO_2 crystallisation, reduces surface roughness, and improves wear as well as electrochemical performance. Coatings deposited at a 50% Ar/O_2 ratio exhibited the best overall properties, including low friction, minimal material loss, and high polarisation resistance, due to the formation of a dense, stable passive ZrO_2 layer. These results highlight the potential of optimised Ag-ZrO_2 coatings for biomedical applications, particularly in implants, and underscore the need for long-term in vitro and in vivo studies to assess their stability and biocompatibility [17].

In their work, Huo et al. report a scalable and efficient strategy to enhance the corrosion resistance of nickel coatings through precise microstructural design, eliminating the need for complex alloying. Using a one-step electrodeposition approach, nanocrystalline Ni coatings with tunable heterogeneous grain structures were systematically developed by controlling the applied current density. An optimal heterostructure, obtained at 10 mA/cm^2, exhibits the best corrosion performance in a 3.5 wt.% NaCl solution, characterised by the

https://doi.org/10.3390/coatings16040407

lowest corrosion current density and the highest charge-transfer resistance. This superior behaviour is attributed to the synergistic interaction between fine and coarse grains, which effectively redistributes chloride adsorption sites and facilitates the formation of a stable and protective NiOOH passive layer. Through a combined approach involving electrochemical analysis, advanced microscopy, and molecular dynamics simulations, the study elucidates the fundamental corrosion mechanisms. Grain boundaries are identified as preferential sites for Cl^- adsorption; however, their negative impact is mitigated in heterostructure systems by spatially dispersing corrosion activity. Conversely, higher current densities induce grain coarsening, crystallographic texture evolution, and amorphous, porous corrosion product formation, ultimately diminishing protective performance. Importantly, this study underscores the potential of compositional planification, demonstrating that tailored microstructural heterogeneity in pure metals can achieve high performance comparable to, or exceeding, that of more complex alloy systems. The findings provide valuable mechanistic insight and establish a practical framework for the design of durable, cost-effective, and sustainable coatings for demanding offshore and industrial environments [18].

Yang et al. demonstrate that the presence of the nickel interlayer reduced the corrosion current density from and significantly decreased the stress between the coating and the substrate. It successfully suppresses crack initiation at the ceramic coating/Ni interface and promotes the formation of a more stable passive film, as reflected by the improved corrosion potential and reduced corrosion current density. However, the protective capability of the Ni interlayer is constrained by the inferior corrosion resistance of the substrate, leading to stress concentration and crack formation at the Ni/substrate interface. This highlights the critical importance of substrate properties in determining overall system durability. The stress state transition from tensile to compressive along the thickness direction further emphasises the complex mechanical interactions governing coating performance. Moreover, under combined corrosion and mechanical impact conditions, coating degradation is dominated by the synergistic action of through-thickness cracks and corrosion pit formation. Through-thickness cracks serve as the primary pathways for interfacial failure, while corrosion pits originate from progressive layer delamination, accelerating coating breakdown. Overall, the findings underscore the importance of interfacial and microstructural design in achieving durable coating systems. The results provide valuable insight into coupled mechanical–electrochemical degradation mechanisms, establishing a solid foundation for the development of advanced, multi-layered protective coatings for demanding industrial and marine applications [19].

Borko et al. investigated carbon steel AISI 1018, reporting that it is highly prone to corrosion in CO_2-saturated brine, a common challenge in oil production and transportation systems. In their study, they demonstrated cerium chloride ($CeCl_3$) and imidazoline oleic acid (IOA), individually and in combination, as environmentally acceptable corrosion inhibitors at 60 °C. Weight-loss, electrochemical, and surface analyses consistently showed that the $CeCl_3$ -IOA mixtures provided superior inhibition efficiency, reaching up to 97.9%, significantly outperforming individual components. IOA forms an adsorbed surface film that contains voids, while cerium carbonate from $CeCl_3$ fills these voids, resulting in a denser, more protective layer. The combination of 15 $mg{\cdot}dm^{-3}$ IOA and 15 $mg{\cdot}dm^{-3}$ $CeCl_3$ yielded the lowest corrosion rate (0.03–0.05 $mm{\cdot}year^{-1}$) and provided stable long-term protection. Mechanistic insights indicate that IOA primarily acts as an anodic inhibitor, whereas $CeCl_3$ and the mixture function as mixed inhibitors with predominant anodic effects. These findings highlight the synergistic potential of combining adsorption-based and film-forming inhibitors, offering a practical and sustainable strategy for CO_2 corrosion control in carbon steel systems [20].

Ortiz et al. demonstrated that electroless Ni-P coatings significantly improve both the corrosion resistance and mechanical properties of ductile iron. The coatings are uniform, strongly adherent, and predominantly amorphous, leading to a substantial increase in surface hardness. Electrochemical studies confirm that the Ni-P layer effectively blocks electrolyte access, facilitates the formation of protective corrosion products, and stabilises localised corrosion mechanisms. These results indicate that electroless Ni-P coatings offer a practical and scalable solution for extending the service life of ductile iron components in harsh environments, making them highly suitable for automotive, energy, and industrial applications. Ductile iron is widely used in engineering applications due to its favourable combination of strength, toughness, and machinability, yet its susceptibility to corrosion limits its service life. In this study, electroless Ni-P coatings were deposited on ductile iron to enhance corrosion resistance and mechanical performance. Comprehensive characterisation using SEM, EDS, X-ray diffraction, and microhardness testing confirmed that the Ni-P coatings are homogeneous, compact, strongly adherent, and predominantly amorphous, leading to a 53% increase in surface hardness. Electrochemical evaluations in 0.1 M NaCl solution revealed that the Ni-P coatings significantly improve corrosion resistance. EIS and phase angle analyses demonstrated that the coating effectively blocks electrolyte access, promotes the formation of a protective corrosion product layer, and shifts the corrosion mechanism toward more stable and localised processes. The impedance of coated samples increased by an order of magnitude compared to the uncoated substrate, highlighting the protective role of the amorphous Ni-P layer. Equivalent circuit modelling further corroborated these findings, showing the formation and evolution of corrosion product films over prolonged immersion. Overall, the study illustrates that electroless Ni-P coatings represent an effective and industrially scalable approach to mitigate corrosion in ductile iron while simultaneously enhancing mechanical properties. These findings provide valuable insights for extending the lifespan of ductile iron components in demanding environments across automotive, energy, and manufacturing sectors [21].

In their work, Fan et al. introduced a scalable and environmentally benign strategy for the fabrication of coloured superhydrophobic stainless steel surfaces, combining high-temperature oxidation with subsequent nano-silica film deposition. The oxidation process enables controllable surface colouration through optical interference effects, while the nano-silica modification effectively transforms initially hydrophilic oxide layers into highly water-repellent, self-cleaning interfaces. Systematic microstructural and electrochemical analyses demonstrate that oxidation temperature plays a critical role in governing surface morphology, wettability, and corrosion performance. Oxide films formed at intermediate temperatures (~600 °C) exhibit optimal corrosion resistance, whereas higher temperatures lead to increased roughness and hydrophilicity. Notably, the application of the nano-silica coating overcomes these limitations, producing superhydrophobic surfaces with contact angles exceeding 150° and markedly reduced corrosion current densities. Electrochemical results further confirm a substantial enhancement in protective performance, with polarisation resistance increasing by several orders of magnitude and corrosion inhibition efficiency approaching 99%. This improvement is attributed to the synergistic effect of hierarchical surface roughness, air entrapment, and the formation of a dense, adherent barrier that effectively restricts electrolyte penetration. Overall, this study provides a compelling pathway for integrating aesthetic functionality with advanced surface protection, offering significant potential for applications in architecture, decorative engineering, and harsh service environments [22].

In this comprehensive review, Samardžija et al. highlight the key role of scanning electrochemical microscopy (SECM) as a superior analytical technique for investigating degradation and corrosion mechanisms of nanocomposite epoxy coatings. SECM's high spatial

https://doi.org/10.3390/coatings16040407

resolution has been shown to enable detailed, localised electrochemical analysis of surface defects and active sites, providing key insights into how nanoparticle modifications—such as ZrO_2, TiO_2, and graphene oxide—improve coating performance. By examining different modes of operation and emphasising the importance of accurate calibration and data interpretation, the review demonstrates SECM's ability to assess coating integrity under realistic environmental conditions. Its ability to deliver spatially resolved electrochemical information surpasses traditional methods, facilitating a deeper understanding of localised corrosion processes that govern coating durability. Integrating SECM with complementary surface analysis techniques shows promise for accelerating the development of next-generation nanocomposite coatings with superior protective properties. Ultimately, adopting SECM in coating research advances the design of more resilient materials, extending service life and ensuring reliability in aggressive environments [23].

Author Contributions: Conceptualisation, M.K. and I.S.; validation, M.K. and I.S.; resources, M.K. and I.S.; writing—original draft preparation, M.K. and I.S.; writing—review and editing, M.K. and I.S. All authors have read and agreed to the published version of the manuscript.

Funding: This research received no external funding.

Conflicts of Interest: The authors declare no conflicts of interest.

References

1. Abdeen, D.H.; El Hachach, M.; Koc, M.; Atieh, M.A. A Review on the Corrosion Behaviour of Nanocoatings on Metallic Substrates. *Materials* **2019**, *12*, 210. [CrossRef]
2. Aljibori, H.S.; Alamiery, A.; Kadhum, A.A.H. Advances in corrosion protection coatings: A comprehensive review. *Int. J. Corros. Scale Inhib.* **2023**, *12*, 1476–1520. [CrossRef]
3. Honarvar Nazari, M.; Zhang, Y.; Mahmoodi, A. Nanocomposite Organic Coatings for Corrosion Protection of Metals: A Review of Recent Advances. *Prog. Org. Coat.* **2021**, *162*, 106573. [CrossRef]
4. Ghader, M.; Bi, H.; Dam-Johansen, K. Advanced materials for smart protective coatings: Unleashing the potential of metal/covalent organic frameworks, 2D nanomaterials and carbonaceous structures. *Adv. Colloid Interface Sci.* **2024**, *323*, 103055. [CrossRef] [PubMed]
5. Tang, S.; Lei, B.; Feng, Z.; Guo, H.; Zhang, P.; Meng, G. Progress in the Graphene Oxide-Based Composite Coatings for Anticorrosion of Metal Materials. *Coatings* **2023**, *13*, 1120. [CrossRef]
6. Faccini, M.; Bautista, L.; Soldi, L.; Escobar, A.M.; Altavilla, M.; Calvet, M.; Domènech, A.; Domínguez, E. Environmentally Friendly Anticorrosive Polymeric Coatings. *Appl. Sci.* **2021**, *11*, 3446. [CrossRef]
7. Jena, G.; Philip, J. A review on recent advances in graphene oxide-based composite coatings for anticorrosion applications. *Prog. Org. Coat.* **2022**, *173*, 107208. [CrossRef]
8. Cao, X.; Pan, J.; Cai, G.; Hu, Y.; Zhang, X.; Dong, Z. Degradation mechanisms of corrosion and biofouling resistance of superhydrophobic coatings in harsh marine conditions. *Prog. Org. Coat.* **2022**, *173*, 107222. [CrossRef]
9. Priyadharshini, R.; Xavier, J.R. Innovative Graphene-Based Nanocomposite Coatings for Advanced Corrosion Resistance in the Automotive Industry: Materials, Mechanisms, and Applications. *Mater. Corros.* **2026**, 1–39. [CrossRef]
10. Yadav, A.; Panjikar, S.; Singh Raman, R.K. Graphene-Based Impregnation into Polymeric Coating for Corrosion Resistance. *Nanomaterials* **2025**, *15*, 486. [CrossRef]
11. Thomas, P.; Sahoo, B.N.; Thomas, P.J.; Greve, M.M. Recent advances in emerging integrated anticorrosion and antifouling nanomaterial-based coating solutions. *Environ. Sci. Pollut. Res.* **2024**, *31*, 67550–67576. [CrossRef]
12. Muresan, L.M. Nanocomposite Coatings for Anti-Corrosion Properties of Metallic Substrates. *Materials* **2023**, *16*, 5092. [CrossRef]
13. De Luna, M.S. Recent Trends in Waterborne and Bio-Based Polyurethane Coatings for Corrosion Protection. *Adv. Mater. Interfaces* **2022**, *9*, 2101775. [CrossRef]
14. Yadav, L.; Sihmar, A.; Kumar, S.; Dhaiya, H.; Vishwakarma, R. Review of nano-based smart coatings for corrosion mitigation: Mechanisms, performance, and future prospects. *Environ. Sci. Pollut. Res.* **2025**, *32*, 17032–17058. [CrossRef] [PubMed]
15. Alam, M.; Alandis, N.M.; Ahmad, N.; Husain, F.M. Anticorrosive and antibacterial nanocomposite coating material from a sustainable resource. *Ind. Crops Prod.* **2020**, *158*, 112955. [CrossRef]
16. Šnajdar, M.; Kurtela, M.; Ćorić, D.; Sakoman, M. Electrochemical Corrosion Performance of TiN, TiCN and TiBN Multilayer Coatings on Hardmetal Substrates. *Coatings* **2026**, *6*, 353. [CrossRef]

https://doi.org/10.3390/coatings16040407

17. Aperador, W.; Orozco-Hernández, G. Improvement in the Corrosion and Wear Resistance of ZrO_2-Ag Coatings on 316LVM Stainless Steel Under Tribocorrosive Conditions. *Coatings* **2025**, *15*, 862. [CrossRef]
18. Huo, W.; Zhang, Z.; Huang, X.; Wang, Y.; Wang, S.; Lu, X.; Li, S.; Zhu, S.; Fang, F.; Jiang, J. Tuning Nanocrystalline Heterostructures for Enhanced Corrosion Resistance: A Study on Electrodeposited Ni Coatings. *Coatings* **2025**, *15*, 534. [CrossRef]
19. Yang, H.; Zhao, Y.; Qin, X.; Jin, Y.; Zhao, X.; Tian, K.; Wang, X.; Zhang, Z. The Damage Evolution of a Cr_2O_3-TiO_2 Coating Subjected to Cyclic Impact and Corrosive Environments and the Influence of a Nickel Intermediate Layer. *Coatings* **2025**, *15*, 98. [CrossRef]
20. Borko, T.; Bilić, G.; Žbulj, K.; Ćurković, H. Individual and Joint Effect of Oleic Acid Imidazoline and $CeCl_3$ on Carbon Steel Corrosion in CO_2-Saturated Brine Solution. *Coatings* **2025**, *15*, 93. [CrossRef]
21. Ortiz, N.; González-Parra, J.; Olaya, J.; Agredo, D.; Valdez, R.; Waage, H.; Bolarin, A.M.; Sánchez, F.; Barba-Pingarrón, A. Morphological and Corrosion Characterisation of Electroless Ni-P Coatings Deposited on Ductile Iron. *Coatings* **2024**, *14*, 1317. [CrossRef]
22. Fan, C.; Wang, X.; Wang, W.; Meng, D.; Zhan, X.; Yin, X.; Liu, Y. Fabrication and Properties of Superhydrophobic Colored Stainless Steel Surface for Decoration and Anti-Corrosion. *Coatings* **2024**, *14*, 1117. [CrossRef]
23. Samardžija, M.; Kurtela, M.; Stojanović, I.; Alar, V. Scanning Electrochemical Microscopy for Investigating Nanocomposite Epoxy Coating Degradation and Corrosion Mechanisms. *Coatings* **2026**, *16*, 165. [CrossRef]

Disclaimer/Publisher's Note: The statements, opinions and data contained in all publications are solely those of the individual author(s) and contributor(s) and not of MDPI and/or the editor(s). MDPI and/or the editor(s) disclaim responsibility for any injury to people or property resulting from any ideas, methods, instructions or products referred to in the content.

https://doi.org/10.3390/coatings16040407

coatings

MDPI

Article

Fabrication and Properties of Superhydrophobic Colored Stainless Steel Surface for Decoration and Anti-Corrosion

Changfeng Fan [1], Xue Wang [1,2,*], Wei Wang [1], Dechao Meng [1], Xianghua Zhan [1], Xiaoli Yin [1] and Yancong Liu [3]

1 Shandong Institute of Petroleum and Chemical Technology, Dongying 257061, China
2 College of Chemical Engineering, China University of Petroleum, Qingdao 266580, China
3 College of Mechanical and Electronic Engineering, China University of Petroleum, Qingdao 266580, China
* Correspondence: xuewang0822@163.com

Abstract: A colored superhydrophobic surface on a stainless steel substrate was achieved by means of high temperature oxidation combined with subsequent spraying modification by superhydrophobic nano-silica film. Comprehensive characterizations of the surface were performed in terms of color, morphology, composition, wettability, and corrosion resistance by optical microscopy, X-ray diffraction (XRD), scanning electron microscopy (SEM), energy-dispersive spectroscopy (EDS), contact angle, potentiodynamic polarization, and electrochemical impedance spectroscopy measurement. At 400 °C, the surface was pale yellow, gradually turning yellow and then red as the temperature increased. At 700 °C and 800 °C, the surface colors were blue and dark brown, respectively. The samples with oxide films demonstrated lower contact angles, specifically 80.5° ± 2.5 at 400 °C, 79.1° ± 2.8 at 500 °C, and 75.6° ± 3.4 at 600 °C. The polarization resistance measured on the oxidized film formed at 600 °C exceeded 7.93×10^4 Ω·cm^2. After spraying the treatment, these colorful surfaces exhibited superhydrophobicity, they were self-cleaning, and they satisfied anti-corrosion properties. The treatment performs as an excellent barrier and exhibits a high corrosion resistance of 4.68×10^6 Ω·cm^2. The successful preparation of superhydrophobic colored surfaces offers the possibility of providing stainless steel with both decoration value and self-cleaning function simultaneously by our proposed chromium-free fabrication process.

Keywords: colored stainless steel; superhydrophobic nano-silica film; contact angle; electrochemical measurement; anti-corrosion properties

1. Introduction

The increasing demand for colored stainless steel is particularly evident in architectural design and decoration due to its attractive appearance, corrosion resistance, and reusability. The color of the stainless steel surface is a consequence of the light interference effect, which is due to the interference amplification of reflected light at both the steel film/oxide and oxide film/air surfaces [1–3].

The reported coloring technologies mainly focus on the production process of H_2SO_4 solutions containing CrO_3 [4,5]. Chromium-based composite coatings are commonly employed as a means of corrosion protection for metals. Alternatives to chromium coatings are becoming more necessary due to their high toxicity and environmental risks. Various methods have been used to produce colored stainless steel surfaces, including thermal oxidation in air [6,7], chemical or electrochemical oxidation in an electrolyte solution [8], and laser machining [9,10].

With its high surface energy, the colored film composed of metal oxides and hydroxides is a porous coating that can absorb liquid pollutants and particles as well as other particles. Therefore, it is necessary to maintain the aesthetic qualities and corrosion resistance of colored stainless steel through frequent manual cleaning. Additionally, physical cleaning methods can sometimes damage the color film. In contrast, superhydrophobic surfaces, with low surface energy and micro-scale roughness, have self-cleaning properties [11–14].

Thus, superhydrophobic modification can be employed to address the issues encountered by colored stainless steel surfaces. The surface becomes less effective in absorbing contaminants after being modified, which prevents the safe contact of the corrosive environment with the surface for extended periods. In addition, superhydrophobic stainless steel offers unique properties such as self-cleaning, drag reduction, anti-icing, and anti-adhesion [15–18], expanding its potential applications in specialized fields.

Surface protection can be achieved through the use of ceramic composite coatings, such as transitional metal oxides like iron, nitrides, carbides, or silicides. There has been a lot of research and development undertaken to modify conventional stainless surfaces to become superhydrophobic [19,20]. Deng et al. prepared the superhydrophobic nano-ZnO (CSS–ZnO) surface on its surface by employing the hydrothermal method. Different structural morphologies of 304 stainless steel-based ZnO surfaces were prepared by varying different seed layer preparation methods [21]. A colored super-hydrophobic surface on stainless steel substrate was achieved by means of chemical coloring on etched substrate combined with modification by fluorosilane. The effect of chemical etching with HCl and $FeCl_3$, and of passivation treatment for the etched substrate by H_2O_2/HNO_3 on coloring behavior, was studied [22]. The effects of NaCl and $NaNO_3$ on the surface morphology and wettability of stainless steel were investigated in detail using various electrolyte concentrations, current densities, and etching time, and the results are expected to further enlarge the application range and improve the service life of 304 stainless steel [23]. Wire electrochemical etching with neutral sodium nitrate is adopted for the first time to fabricate superhydrophobic surfaces on 304 stainless steel substrates [24]. The prepared superhydrophobic 304 stainless steel surfaces show excellent anti-icing performance and self-cleaning properties, which may greatly facilitate the application of 304 stainless steel in extreme conditions. The development of corrosion inhibitors that also impart hydrophobic properties to metallic surfaces has been an area of significant interest. For instance, the use of alkanethiols on gold surfaces, as demonstrated by Renner and colleagues [25], has been shown to effectively render the surface hydrophobic while inhibiting corrosion through a mechanism involving spinodal decomposition dealloying. Similarly, a superhydrophobic nano-silica coating on colored stainless steel not only enhances the surface's resistance to wetting but also provides a protective barrier against corrosive agents. The nano-silica coating, with its high-temperature stability and corrosion resistance, offers a comparable level of protection, albeit through a different mechanism involves the formation of a uniform and adherent film that prevents corrosive species from reaching the metal surface. There are various methods for preparing silica coatings, the main ones being the sol-gel process and the hydrothermal method. In particular, the sol-gel process offers advantages for preparing silica coatings on stainless steel substrates, including lower synthesis temperature, easier control of reaction conditions, and simpler equipment and process requirements. The surface can be coated using spraying, which is a simple process that can also be scaled and improved. This method can overcome the disadvantages of the high-temperature hydrothermal method including the susceptibility to corrosion of stainless steel. Silica film improves the anti-corrosion properties and hydrophobicity of stainless steel surfaces. However, the study of superhydrophobically colored stainless steel surfaces fabricated by high temperature oxidation has yielded minimal results.

In this study, we present a novel method for fabricating superhydrophobic colored stainless steel surfaces by high temperature oxidation combined with the application of superhydrophobic nano-silica film spraying. The surface characterization of the oxide films was performed by optical microscopy and X-ray diffraction. Scanning electron microscopy and energy-dispersive spectroscopy were employed to investigate the chemical composition of the sprayed films. The corrosion properties were studied using potentiodynamic polarization and electrochemical impedance spectroscopy. This analysis can contribute to the development of heat treatment processes for high temperature applications or for requirements regarding color and corrosion resistance. The successful preparation of superhydrophobic colored surfaces offers the opportunity to give stainless steel both decorative value and self-cleaning functionality through our proposed chromium-free manufacturing process.

2. Experimental Section

2.1. Sample Preparation

The material applied was a commercial 304 stainless steel with the chemical composition listed in Table 1. Hexachloroplatinic acid, isopropanol, Triethoxysilane ($Si(CH_3CH_2O)_3H$), Anhydrous ethanol, tetraethyl orthosilicate (TEOS), and ammonia were purchased from Alfa Aesar. Allyl polyethylene glycol polypropylene methyl ether (AEPM) was obtained from Shanghai Kelaman Chemical Technology Co., Ltd., Shanghai, China.

Table 1. Chemical composition of 304 stainless steel used (in wt%).

Element	S	P	C	Mo	Cu	Si	Mn	Ni	Cr	Fe
Content	0.007	0.027	0.057	0.15	0.23	0.59	1.05	8.09	18.14	Bal.

Samples with dimensions of 30 mm × 30 mm × 1 mm were cut from the steel plate with an initial surface roughness that adheres to industrial standards, within the range of Ra 0.4 to 0.8 μm. The samples were treated with acetone for ultrasonic cleaning and dried under hot air. Samples were subjected to oxidation in dry air with a dew point low enough to prevent moisture from affecting the coating process for 30 min at 400 °C, 500 °C, 600 °C, 700 °C, and 800 °C, respectively. The samples were cooled within the furnace to room temperature after the heat treatment was completed.

2.2. Fabrication of Nano-Silica Particle

The Speier catalyst solution was prepared by dissolving 1 g of hexachloroplatinic acid in 20 mL of pure isopropanol, which served as the solvent. The solution was then diluted 100-fold with anhydrous isopropanol to ensure uniform distribution of the catalyst. A predetermined quantity of polypropylene polyethylene glycol methyl ether (AEPM) was mixed with the diluted catalyst in a nitrogen atmosphere to prevent any unwanted side reactions. Triethoxysilane ($Si(CH_3CH_2O)_3H$) was then gradually added to the mixture and the reaction was allowed to proceed at a controlled temperature of 40 °C for 12 h. The desired coupling agent product was obtained by vacuum distillation, which eliminated the unreacted starting materials and by products from the reaction shown in Figure 1. Figure 2 presents the preparation of hydrophobic nano-silica. Anhydrous ethanol and water were combined in a volume ratio of 95:5 in a three-necked flask. A measured amount of tetraethyl orthosilicate (TEOS) was introduced into the ethanol-water mixture with continuous stirring at room temperature. The mixture was stirred after sealing, and the organosilane was added at a rate of 3 mL per 100 mL of the ethanol-water solution. The pH of the solution was carefully adjusted to 13 with ammonia solution. The reaction mixture was then heated to 40 °C and continued to be stirred magnetically for 8 h. This process facilitated the in-situ synthesis of hydrophobically modified nanosized silica.

The silica suspension was loaded into a spray bottle and sprayed onto the colored stainless steel surface at a speed of 3–5 cm/s with the nozzle approximately 15 cm away from the surface. After that, the sprayed surface was subjected to heat treatment in an oven at 120 °C for 30 min to create a superhydrophobic coating of stainless steel.

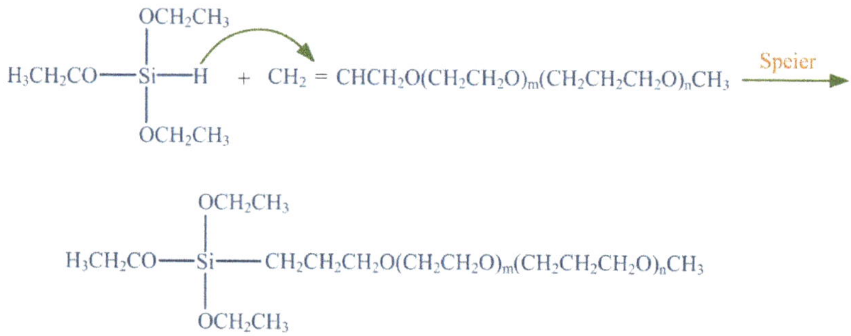

Figure 1. The preparation of organosilane coupling agents.

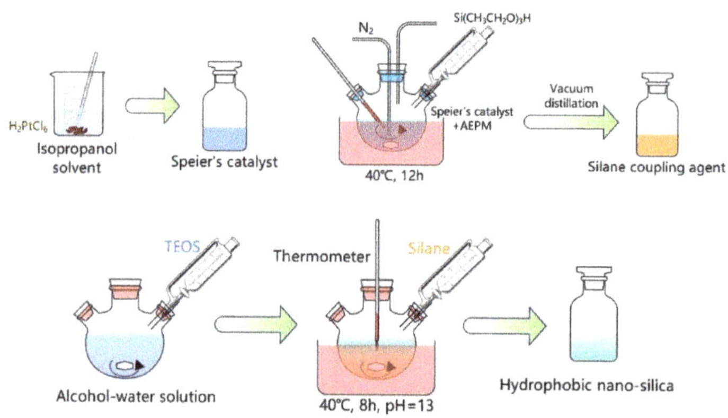

Figure 2. The preparation of hydrophobic nano-silica.

2.3. Sample Characterization

Scanning electron microscopy (SEM, S3400/SU3800, Hitachi, Tokyo, Japan) was used to demonstrate the surface morphology of stainless steel. X-ray diffraction (XRD: SHIMADZU 6100, Kyoto, Japan) was used to analyze the surface chemistry, while energy dispersive spectroscopy (EDS, Xplore30, Oxford Instruments, Oxfordshire, UK) was used to characterize the surface chemistry on the coated stainless steel.

2.4. Contact Angle Measurements

The water repellency of the samples was assessed by determining the static contact angle with the sessile drop method, utilizing a video-based optical contact angle meter (JC2000D3M, Zhongchen, Shanghai, China). An 8 μL drop of deionized distilled water was placed on the coated stainless steel surface in ambient conditions.

2.5. Electrochemical Measurements

The electrochemical measurements were performed in a 3.5 wt.% NaCl solution using an electrochemical workstation (Reference 3000, Gamry, Warminster, PA, USA). A silver/silver chloride electrode (Ag/AgCl, 3M KCl) served as reference electrode and a platinum mesh was used as the counter electrode. The working electrode was clamped to a flat cell with an exposure area of 1.0 cm^2 to the electrolyte. The open circuit potential (OCP) was measured for 900 s to characterize the electrochemical response of the samples and ensure the stability of the system. Electrochemical impedance spectroscopy (EIS) measurements were performed at the OCP over a frequency range of 100 kHz to 100 mHz with a signal amplitude of 20 mV. Zview software (Version 3.1, Scribner Associates, Inc.,

Southern Pines, NC, USA) was used to fit and analyze the measurement results with specified equivalent circuit. Potentiodynamic polarization (DP) measurements were conducted with a scanning rate of 1 mV/s over a voltage range of ±0.1 V vs. OCP. Three parallel tests were performed under the same conditions for each sample.

3. Results and Discussion

3.1. Optical Micrographs of Oxidized Stainless Steel Surfaces

Optical micrographs of colored stainless steel formed at different oxidation temperatures are presented in Figure 3. Visible color changes were observed on the stainless steel surface at different oxidation temperatures. At 400 °C, the surface was pale yellow, gradually turning yellow and then red as the temperature increased. The surface after a short period of operation is rich in iron. As the temperature increases, a combination of Cr_2O_3, spinel $(Fe,Cr)_3O_4$ and iron oxide can form in the oxide film [26]. At 700 °C and 800 °C, the surface colors are blue and dark brown, respectively. Studies on the coloring mechanism of stainless steel have demonstrated that the coloration is a result of the interference of natural light reflected from different optical path differences within the oxide film [7]. This interference effect is influenced by the thickness and the refractive index of the oxide layer, which alters the way light interacts with the surface. The formation of a transparent oxide film during the oxidation process creates a multilayer structure that contributes to the coloration effect observed in our samples. Film thickness is an important factor in coloring. At low temperatures the oxide film is thinner, while it becomes thicker at high temperatures. When the coloring temperature exceeds 800 °C, the oxide particles become coarse, forming a spinel structure, resulting in a rough surface. This rough surface causes diffuse reflection, reducing brightness. The surface morphology and thickness change with the formation of the oxide, further affecting the corrosion resistance. The superhydrophobic nano-silica film did not affect the color of the high-temperature oxidized stainless steel surface.

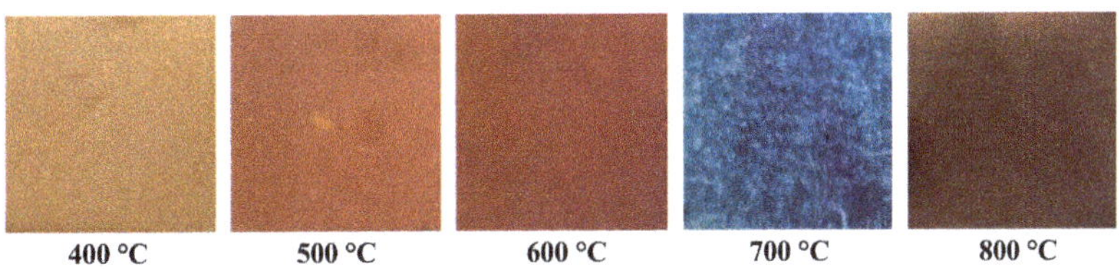

Figure 3. Optical micrographs of oxide films formed at different oxidation temperatures.

The oxide film itself is a transparent layer, with the observed colors resulting from interference effects caused by varying optical path lengths of reflected light on the oxide surface, influenced by natural light as mentioned above. When incident light strikes the oxide film perpendicularly, the optical path difference between reflected and refracted light is given by $2nd$, where n is the refractive index of the oxide film and d is the film thickness. Interference occurs when this optical path difference is an integer multiple of the wavelength λ of the light.

$$k\lambda = 2nd \ (k = 0,1,2,3\ldots) \quad (1)$$

Thus, the film thickness d can be calculated using the following formula:

$$d = k\lambda/2n \ (k = 0,1,2,3\ldots) \quad (2)$$

where for stainless steel films, the refractive index n ranges from 2.64 to 2.73 [7]. The thickness of the oxide film for gold-colored stainless steel ranges from 135.5 to 140.2 nm. In contrast, the thickness of the oxide film for blue-violet stainless steel ranges from 512.8 to 530.3 nm.

3.2. Surface Characterization

Figure 4 shows the XRD patterns of 304 stainless steel oxide films formed at different temperatures for 30 min. XRD measurements of these representative samples provide the basis for further investigation of their chemical composition. α-Fe is present in samples formed at temperatures below 600 °C. As the oxidation temperature rises, the primary oxides formed on the sample surface are the spinel M_3O_4 and Cr_2O_3 phases. The relative intensities of the oxide peaks are significantly lower than those of the substrate, due to the low phase composition and/or the thin film thickness. The M element corresponds to Cr and/or Fe. The exact composition of spinel cannot be determined by simply observing the positions of its peaks in the XRD pattern. The ratio of Cr_2O_3 and Fe_3O_4 in the samples significantly affects their corrosion resistance, providing an indication of their corrosion resistance to some extent.

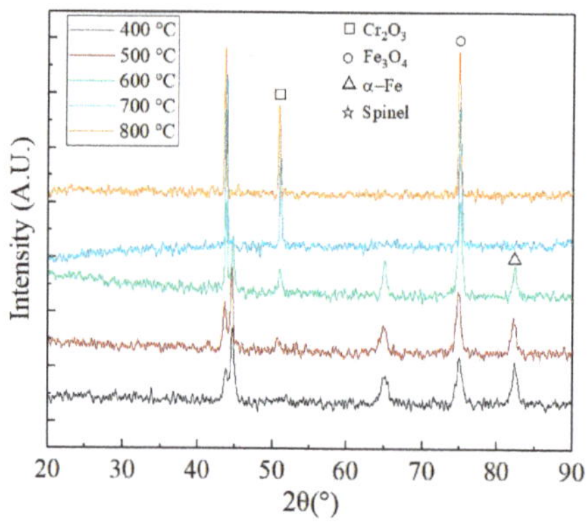

Figure 4. X-ray diffraction patterns of oxide film formed at different temperatures on 304 stainless steel.

Figure 5 shows typical SEM micrographs of oxide films formed at 800 °C and films sputtered with superhydrophobic nano-silica films. As shown in Figure 5a, the surface of the oxide film formed at 800 °C is very rough, characterized by a dense coverage of small epitaxial oxide particles and larger spinel oxides. This rough surface texture leads to diffuse reflection, which reduces the overall brightness of the surface. In addition, this roughness also enhances wettability, making the surface more susceptible to water adhesion. The oxide film formed at 800 °C mainly consists of spinel M_3O_4 and Cr_2O_3 phases. This composition is due to the fact that at temperatures above 600 °C, the diffusion rate of chromium is high enough to transport chromium atoms to the surface. Once on the surface, these chromium atoms react with oxygen to form chromium oxide. The specific chemical composition and thickness of these oxide films play an important role in determining their anti-corrosion properties. Then, when the sample surface was sprayed with a layer of superhydrophobic nano-silica film, the SEM image showed a remarkable transformation. The coating generally appears smoother and more uniform than an untreated oxide film. This nano-silica coating forms a dense, adherent film on the surface, improving the hydrophobic properties of the material. However, during the high-temperature processing required to cure the coating, thermal stress can cause micron-scale cracks to form on the film surface. These micro-cracks may form due to a combination of factors, including the differential thermal expansion between the underlying oxide layer and the superhydrophobic coating, as well as the difference in elastic modulus between the two materials. The mismatch in the elastic properties can create mechanical stress at the interface, which, when combined with the thermal stress during cooling, can lead to cracking. Despite these micro-cracks, the overall integrity and performance of the superhydrophobic coating is maintained, providing significant improvements in both hydrophobicity and corrosion resistance.

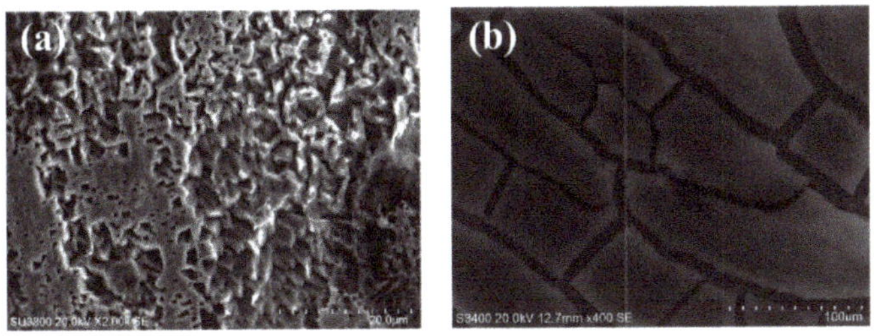

Figure 5. SEM images of (**a**) oxide films formed at 800 °C and (**b**) that are sprayed with a superhydrophobic nano-silica film.

The elemental distribution of the sample oxidized at 800 °C for 30 min and then treated with the superhydrophobic nano-silica film is shown in Figure 6. The EDS mapping results clearly depict the elemental distribution across the surface. In particular, the presence of silicon and oxygen elements confirms the formation of the nano-silica film, which is indispensable for the observed hydrophobic properties and enhances the corrosion resistance of the treated surface. The nano-silica film creates a protective barrier that provides these beneficial properties. However, the EDS mapping also shows the presence of iron, chromium, nickel, and oxygen elements under the superhydrophobic coating, which can be seen through the micro-cracks in the nano-silica film. Observation of these features through micro-cracks shows that although the nano-silica film significantly improves the surface properties, it does not completely mask the underlying oxidation layer. High-temperature oxidation results in the formation of complex oxides consisting of Fe, Cr, and Ni that are characteristic of the protective oxide layer on stainless steel. These oxides might contribute to the intrinsic corrosion resistance of the material, which will be further enhanced by the addition of a superhydrophobic nano-silica film.

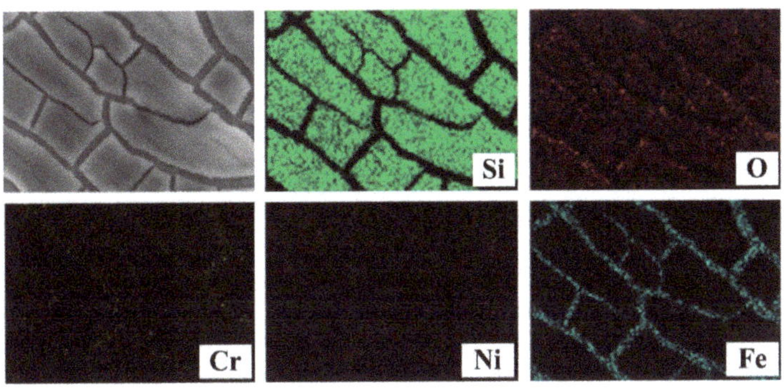

Figure 6. Element distribution of sample oxidized at 800 °C for 30 min and treated with the superhydrophobic nano-silica film.

3.3. Contact Angle Measurements

Figure 7 illustrates the contact angles of the oxide films formed at different oxidation temperatures and the oxide film formed at 800 °C treated with the superhydrophobic nano-silica film. Prior to any treatment, the as-received samples of 304 stainless steel exhibited a contact angle of 91.2° ± 3.1, indicating moderate hydrophobicity. In contrast to the untreated substrate, the samples with oxide films demonstrated lower contact angles, specifically 80.5° ± 2.5 at 400 °C, 79.1° ± 2.8 at 500 °C, and 75.6° ± 3.4 at 600 °C. This reduction in contact angle indicates a transition towards increased hydrophilicity with rising oxidation temperatures. As the oxidation temperature continued to increase, the surface contact angle showed a significant further decrease. At 800 °C, the contact angle reached 46.2° ± 2.2, reflecting a substantial enhancement in surface wettability. This pronounced decrease in contact angle is strongly correlated with the increased surface roughness observed in Figure 5, which enhances the surface's hydrophilic properties. However, when the oxide film formed at 800 °C was treated with the superhydrophobic nano-silica film, a remarkable change in wettability was observed. The contact angle of the treated sample reached 151.5° ± 1.2, indicating excellent hydrophobicity. The application of the superhydrophobic nano-silica film significantly altered the surface characteristics, transforming it from a hydrophilic state to a hydrophobic state. This dramatic increase in contact angle demonstrates the effectiveness of the superhydrophobic nano-silica film in modifying the surface wettability of the oxidized stainless steel. The nano-silica coating creates a uniform and dense film that imparts superhydrophobic properties to the surface. This transformation not only improves water repellence but also enhances the resistance to corrosion and environmental degradation.

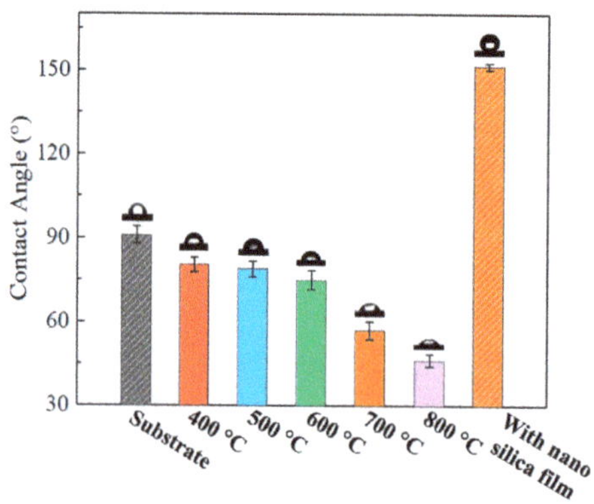

Figure 7. Contact angle of oxide films formed at different oxidation temperatures and oxide film formed at 800 °C treated with the superhydrophobic nano-silica film.

3.4. Dynamic Polarization and Electrochemical Impedance Spectroscopy Measurements

Figure 8a illustrates the corrosion behavior of oxide films formed at different temperatures for 30 min in a 3.5 wt.% NaCl solution. The films exhibit similar corrosion behavior, with slight variations in current densities. These differences can be further analyzed in conjunction with EIS measurement results. The presence of chromium in the oxide films enhances their stability and corrosion resistance, as supported by references [27–29]. Notably, the oxide film formed at 600 °C demonstrates the lowest corrosion current density 0.62 $\mu A/cm^2$, indicating superior anti-corrosion performance. As the oxidation temperature increases, the current density also increases, reaching a peak at 800 °C. This reduction in corrosion resistance at higher temperatures is attributed to the increasing content of Fe_3O_4, which leads to a corresponding decrease in the proportion of Cr_2O_3 in the coating. Figure 8b compares the dynamic polarization curves of the substrate, the oxide films formed at 800 °C, and those treated with the superhydrophobic nano-silica film. Compared to the oxide film formed at 800 °C, a significant decrease in current density is observed for the sample treated with the superhydrophobic nano-silica film, dropping from 4.13 $\mu A/cm^2$ to 0.029 $\mu A/cm^2$. This substantial reduction highlights the effectiveness of the superhydrophobic nano-silica film in isolating the substrate from corrosive media, thereby significantly enhancing its anti-corrosion performance. The superhydrophobic coating acts as a barrier, preventing the penetration of corrosive agents and maintaining the integrity of the underlying oxide film. This treatment not only improves corrosion resistance but also extends the longevity of the stainless steel surface in harsh environments.

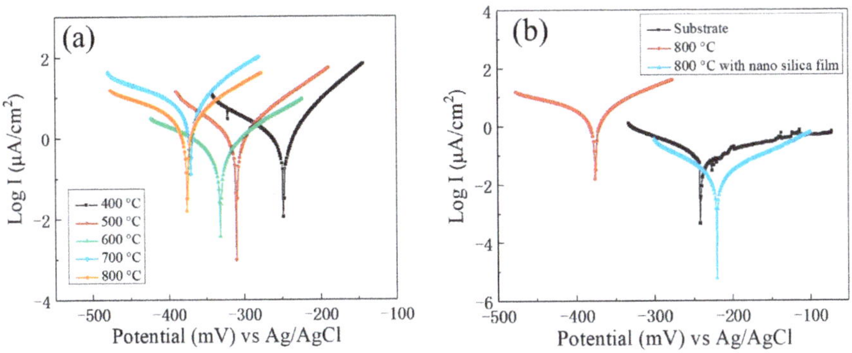

Figure 8. Dynamic polarization curves of oxide films in 3.5 wt% NaCl: (**a**) oxide film formed at different temperatures for 30 min; (**b**) substrate and oxide film formed at 800 °C treated with the superhydrophobic nano-silica film.

Electrochemical impedance spectroscopy (EIS) was utilized to evaluate the corrosion performance of the oxide films. The Nyquist plots of oxide films on stainless steel formed at different temperatures for 30 min are presented in Figure 9a. The spectra exhibit similar shapes with two time constants for the oxide films at high and low frequencies, as evidenced by the depressed semi-circles for the oxidized samples. In contrast, the Nyquist plot of the substrate displays a capacitive arc that does not end in the low-frequency range. The oxide films formed at 400–600 °C show higher impedance values of 6.93×10^4, 7.48×10^4, and 7.93×10^4 $\Omega \cdot cm^2$, respectively. When the oxidation temperature exceeds 600 °C, the impedance decreases, consistent with the dynamic polarization results.

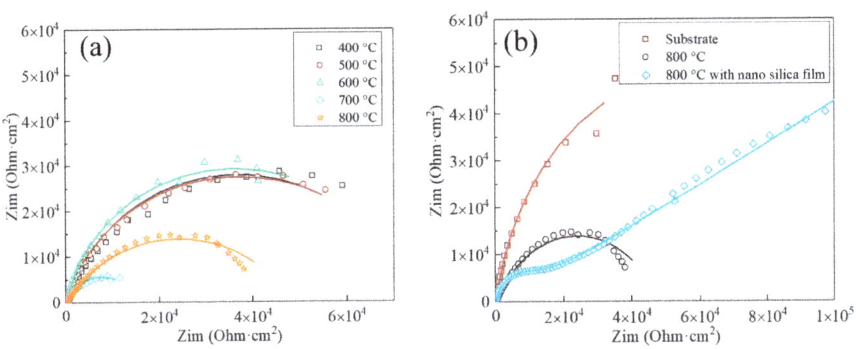

Figure 9. Nyquist plots of oxide films in 3.5 wt% NaCl: (**a**) oxide film formed at different temperatures for 30 min; (**b**) substrate and oxide film formed at 800 °C treated with the superhydrophobic nano-silica film.

Figure 9b displays the Nyquist plots of the substrate and the oxide films formed at 800 °C treated with the superhydrophobic nano-silica film. The EIS results reveal one relaxation process for the substrate and two for the oxide film treated with the superhydrophobic nano-silica film. Compared to the samples with oxide films, the substrate exhibits higher impedance, which is likely due to the presence of a dense passive film on the substrate surface. This dense passive film reduces the electrochemical reaction rates on the surface of the stainless steel, whereas the oxide films impair the corrosion resistance.

The equivalent circuit depicted in Figure 10a has been applied to model the interface between the electrolyte and the electrode for a substrate with a passive film. The equivalent circuit consists of the electrolyte resistance (R_e) connected in series with one time constant.

R_f and Q_f represent the resistance and capacitance behavior of the passive film, respectively. The constant phase element (CPE) is introduced to account for the dispersion behavior resulting from various physical phenomena, such as local inhomogeneity in the dielectric material and surface roughness. The CPE represents a deviation from a pure capacitor [30].

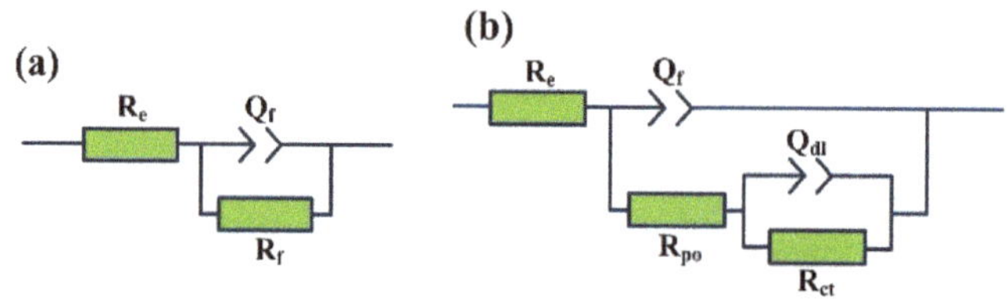

Figure 10. Equivalent circuits used to interpret electrochemical impedance spectra of (**a**) substrate and (**b**) oxide films formed at different temperatures for 30 min and treated with the superhydrophobic nano-silica film when exposed to 3.5 wt% NaCl solution.

Different models have been proposed to interpret the impedance spectra of oxidized stainless steel [31]. In these models, Q_f represents the capacitive behavior of the oxide film formed, which is coupled with the pore resistance (R_{po}) due to the ionic pathways or microcracks through the film. Q_{dl} indicates the double layer capacitance and R_{ct} stands for the corresponding charge transfer resistance.

Polarization resistance (R_{pl}), defined as $R_{pl} = R_{po} + R_{ct}$, is commonly considered a key parameter to evaluate corrosion-induced deterioration in oxidized samples. A high R_{pl} value implies excellent corrosion prevention capability, whereas a lower value indicates accelerated corrosion due to interaction between the sample and electrolyte [31].

The polarization resistance measured on the oxidized film formed at 600 °C for 30 min is higher and exceeds $7.93 \times 10^4\ \Omega\cdot cm^2$, which is consistent with the observed low corrosion current density. This result can be explained by the fact that oxide films enriched in chromium oxide and spinel can impede the charge transfer process, thereby exhibiting higher corrosion resistance compared to oxide films mainly composed of iron oxide [32]. Conversely, an oxide film formed at 700 °C displays the lowest $R_{pl,\,700\,°C}$, at $1.51 \times 10^4\ \Omega\cdot cm^2$. However, the sample treated with the superhydrophobic nano-silica film presents excellent corrosion performance, with $R_{pl,\,sprayed}$ values reaching $4.68 \times 10^6\ \Omega\cdot cm^2$. The corrosion inhibition efficiency (IE%) based on the polarization resistance values obtained from the electrochemical measurements for both non-sprayed and sprayed samples annealed at 800 °C can be calculated as [33]:

$$IE\% = (1 - R_{pl,\,800\,°C}/R_{pl,\,sprayed}) \times 100\% = 98.97\% \tag{3}$$

These findings suggest that further investigations are essential to fully elucidate the underlying mechanisms. Several factors may contribute to the observed improvements in corrosion resistance and superhydrophobicity. The formation of a uniform nano-silica coating likely creates a robust barrier that minimizes direct contact between the stainless steel surface and the corrosive medium, thereby reducing the electrochemical activity and slowing down the corrosion process. Additionally, the nano-silica film may alter the surface energy distribution, promoting air entrapment within the micro- and nano-structures on the surface, which is known to enhance the superhydrophobic effect. The increased surface roughness post-treatment could also enhance the mechanical interlocking of the nano-silica film, further improving its adherence and protective qualities. The enhanced corrosion resistance of the colored samples compared to the substrate suggests that the superhydrophobic colored film provides a protective effect for the substrate. This improvement highlights the potential of superhydrophobic treatments in extending the durability and corrosion resistance of colored stainless steel surfaces.

As shown in Figure 11a, the oxide film alone results in a rough, colored stainless steel surface that exhibits strong wettability. When exposed to a corrosive medium, the rough structures coated with the superhydrophobic nano-silica film become filled with entrapped air. Consequently, the corrosive medium only contacts the surface at the tips of the micro-protuberances. Additionally, even in these tip contact areas, the superhydrophobic nano-silica film and the colored oxide film act as double barriers, preventing the metal substrate from direct contact with the corrosive medium to a certain extent. Hence, the overall corrosion resistance is greatly reduced. Furthermore, a comparison of oxide films formed at different temperatures reveals that the films display various colors corresponding to different thicknesses. This indicates that thickness alone is not the only factor affecting the corrosion performance of the colored films. Other factors, including composition and microstructural changes, must be considered comprehensively. Therefore, the correlation between corrosion resistance and coloring parameters presents another important scientific issue that warrants further investigation in future studies.

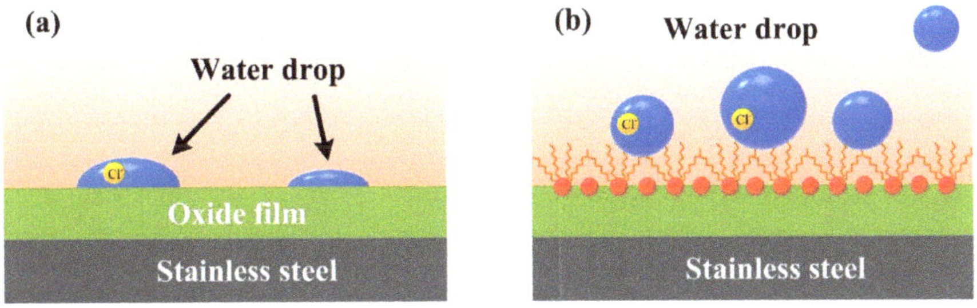

Figure 11. Wetting phenomena of different samples when exposed to corrosive solution: (**a**) oxide films formed at different temperatures for 30 min and (**b**) oxide film treated with the superhydrophobic nano-silica film.

4. Conclusions

In summary, we fabricated colored stainless steel surfaces with different colors successfully by using high temperature oxidation combined with the superhydrophobic nano-silica film. At 400 °C, the surface was pale yellow, gradually turning yellow and then red as the temperature increased. At 700 °C and 800 °C, the surface colors are blue and dark brown, respectively. The samples with oxide films demonstrated lower contact angles, specifically $80.5° \pm 2.5$ at 400 °C, $79.1° \pm 2.8$ at 500 °C, and $75.6° \pm 3.4$ at 600 °C. The polarization resistance measured on the oxidized film formed at 600 °C for 30 min was higher and exceeded 7.93×10^4 $\Omega{\cdot}cm^2$, which is in accordance with the observed low corrosion current density 0.62 $\mu A/cm^2$. The superhydrophobic nano-silica film was synthesized and sprayed on the oxide films formed at different temperatures. Its EIS analysis showed two relaxation process, which performed as an excellent barrier and exhibited high corrosion resistance 4.68×10^6 $\Omega{\cdot}cm^2$. After being sprayed, the colored surface possessed superhydrophobic properties and better corrosion resistance compared to the colored stainless steel surface with the oxide film and substrate. Therefore, our approach offers an environmentally friendly method towards the development of colored superhydrophobic stainless steel surface with both decoration value and self-cleaning function, and its promising applications are anticipated in various indoor or outdoor fields, especially in the design and architectural sectors.

Author Contributions: Conceptualization, C.F.; methodology, C.F. and X.W.; investigation, C.F., W.W. and D.M.; formal analysis, X.Z.; data curation, X.Y.; writing—original draft preparation, C.F.; writing—review and editing, C.F., X.W. and Y.L.; funding acquisition, C.F., X.Z. and X.Y. All authors have read and agreed to the published version of the manuscript.

Funding: This work is supported by Dongying Science Development Fund (DJ2022006), Dongying Major Science and Technology Innovation Project (Science and Technology Development Guidance Plan) (2022ZD55, 2023ZDJH114) and Shandong Provincial Natural Science Foundation (ZR2022QE186, ZR2021ME180, ZR2023QE139).

Institutional Review Board Statement: Not applicable.

Informed Consent Statement: Not applicable.

Data Availability Statement: Data are contained within this article.

Conflicts of Interest: The authors declare no conflicts of interest.

References

1. Qu, J.; Yu, C.; Nie, C.; Wang, H.; Cao, Z.; Li, Y.; Wang, X. A new environmentally friendly approach to prepare superhydrophobic colored stainless steel surface for decoration, anti-corrosion and self-cleaning. *J. Mater. Sci.* **2020**, *56*, 854–869. [CrossRef]
2. Alliott, G.T.; Higginson, R.L.; Wilcox, G.D. Producing a thin coloured film on Stainless Steels—A review. part 2: Non-electrochemical and Laser Processes. *Trans. IMF* **2023**, *101*, 72–78. [CrossRef]
3. Gaidys, M.; Selskis, A.; Gečys, P.; Gedvilas, M. Stainless steel colouring using burst and BIBURST mode ultrafast laser irradiation. *Opt. Laser Technol.* **2024**, *174*, 110561. [CrossRef]
4. Conrrado, R.; Bocchi, N.; Rocha-Filho, R.C.; Biaggio, S.R. Corrosion resistance of colored films grown on stainless steel by the alternating potential pulse method. *Electrochim. Acta* **2003**, *48*, 2417–2424. [CrossRef]
5. Zhang, J.; Yuan, J.; Qiao, Y.; Cao, C.; Zhang, J.; Zhou, G. The corrosion and passivation of SS304 stainless steel under square wave electric field. *Mater. Chem. Phys.* **2003**, *79*, 43–48. [CrossRef]
6. Sharma, V.C. A comparison of thermal performance of austenitic stainless steel solar absorber plates coloured by chemical and thermal oxidation techniques. *Energy* **1981**, *6*, 133–138. [CrossRef]
7. Huang, Q.; Ang, Y.; Ye, S.; Xiang, X. Color of SUS304 stainless steel in high temperature oxidation. *Met. Funct. Mater.* **2017**, *24*, 39–43.
8. de Oliveira Vasconcelos, K.; Bocchi, N.; Rocha-Filho, R.C.; Biaggio, S.R. An environmentally friendly and practical method for obtaining color on stainless steel by interference. *J. Electrochem. Soc.* **2005**, *152*, 11.
9. Luo, F.; Ong, W.; Guan, Y.; Li, F.; Sun, S.; Lim, G.C.; Hong, M. Study of micro/nanostructures formed by a nanosecond laser in gaseous environments for stainless steel surface coloring. *Appl. Surf. Sci.* **2015**, *328*, 405–409. [CrossRef]
10. Guay, J.-M.; Calà Lesina, A.; Côté, G.; Charron, M.; Poitras, D.; Ramunno, L.; Berini, P.; Weck, A. Laser-induced plasmonic colours on metals. *Nat. Commun.* **2017**, *8*, 16095. [CrossRef]
11. Kim, J.-H.; Mirzaei, A.; Kim, H.W.; Kim, S.S. Facile fabrication of superhydrophobic surfaces from austenitic stainless steel (Aisi 304) by Chemical Etching. *Appl. Surf. Sci.* **2018**, *439*, 598–604. [CrossRef]
12. Zhong, M.; Zhang, Y.; Li, X.; Wu, X. Facile fabrication of durable superhydrophobic silica/epoxy resin coatings with compatible transparency and stability. *Surf. Coat. Technol.* **2018**, *347*, 191–198. [CrossRef]
13. Ye, Y.; Kang, Z.; Wang, F.; Long, Y.; Guo, T.; Chen, D.; Kong, J.; Xu, L. Achieving Hierarchical Structure with Superhydrophobicity and Enhanced Anti-Corrosion via Electrochemical Etching and Chemical Vapor Deposition. *Appl. Surf. Sci.* **2022**, *610*, 155362. [CrossRef]
14. Zhang, Y.; Zhang, Z.; Yang, J.; Yue, Y.; Zhang, H. Fabrication of Superhydrophobic Surface on Stainless Steel by Two-Step Chemical Etching. *Chem. Phys. Lett.* **2022**, *797*, 139567. [CrossRef]
15. Li, L.; Breedveld, V.; Hess, D.W. Creation of superhydrophobic stainless steel surfaces by acid treatments and hydrophobic film deposition. *ACS Appl. Mater. Interfaces* **2012**, *4*, 4549–4556. [CrossRef]
16. Emelyanenko, K.A.; Emelyanenko, A.M.; Boinovich, L.B. Laser Obtained Superhydrophobic State for Stainless Steel Corrosion Protection, a Review. *Coatings* **2023**, *13*, 194. [CrossRef]
17. Tang, Y.; Cai, Y.; Wang, L.; Luo, X.; Wang, B.; Song, Q.; Liu, Z. Fabrication of Superhydrophobic Stainless Steel via Hybrid Femtosecond Laser-Chemical Method with Wear-Resistance and Anti-Corrosion Properties. *Opt. Laser Technol.* **2023**, *164*, 109474. [CrossRef]
18. Wysard, M.; Vasconcelos, R.; de Souza, E.A.; Costa, M.E.H.M.; Camargo, S.S. Very Simple Method to Produce Superhydrophobic Stainless Steel Surfaces at Room Temperature. *Mater. Chem. Phys.* **2023**, *307*, 128203. [CrossRef]
19. Park, B.; Hwang, W. A facile fabrication method for corrosion-resistant micro/nanostructures on stainless steel surfaces with tunable wettability. *Scripta. Mater.* **2016**, *113*, 118–121. [CrossRef]
20. Kim, J.; Sim, S.O.; Park, H.W. Fabrication of durable hydrophobic micropatterns on stainless steel using a hybrid irradiation process. *Surf. Coat. Technol.* **2016**, *302*, 535–542. [CrossRef]
21. Deng, Y.; Xu, F.; Yin, Z.; Xue, M.; Chen, Y.; He, P.; Wu, J.; Ou, J.; Wang, F.; Luo, Y.; et al. Controllable fabrication of superhydrophobic alloys surface on 304 stainless steel substrate for anti-icing performance. *Ceram. Int.* **2023**, *49*, 25135–25143. [CrossRef]
22. Qu, J.; Yu, C.; Cui, R.; Qin, J.; Wang, H.; Cao, Z. Preparation of super-hydrophobic and corrosion resistant colored films on chemically etched 304 stainless steel substrate. *Surf. Coat. Technol.* **2018**, *354*, 236–245. [CrossRef]

23. Sun, Y.; Zhao, D.; Song, J. Electrochemical fabrication of superhydrophobic 304 stainless steel by neutral electrolyte. *J. Mater. Sci.* **2022**, *57*, 20069–20081. [CrossRef]
24. Sun, Y.; Liu, J.; Ming, P.; Zhao, D.; Song, J. Wire electrochemical etching of superhydrophobic 304 stainless steel surfaces based on high local current density with neutral electrolyte. *Appl. Surf. Sci.* **2022**, *571*, 151269. [CrossRef]
25. Renner, F.U.; Ankah, G.N.; Bashir, A.; Ma, D.; Biedermann, P.U.; Shrestha, B.R.; Nellessen, M.; Khorashadizadeh, A.; Losada-Pérez, P.; Duarte, M.J.; et al. Star-Shaped Crystallographic Cracking of Localized Nanoporous Defects. *Adv. Mater.* **2015**, *27*, 4877–4882. [CrossRef]
26. Gardiner, D.J.; Littleton, C.J.; Thomas, K.M.; Strafford, K.N. Distribution and characterization of high temperature air corrosion products on iron-chromium alloys by Raman Microscopy. *Oxid. Met.* **1987**, *27*, 57–72. [CrossRef]
27. Wang, S.; Ma, Q.; Li, Y. Characterization of microstructure, mechanical properties and corrosion resistance of dissimilar welded joint between 2205 duplex stainless steel and 16MNR. *Mater. Design* **2011**, *32*, 831–837. [CrossRef]
28. Niu, W.; Lillard, R.S.; Li, Z.; Ernst, F. Properties of the passive film formed on interstitially hardened AISI 316L stainless steel. *Electrochim. Acta* **2015**, *176*, 410–419. [CrossRef]
29. Yu, F.; Chen, S.; Chen, Y.; Li, H.; Yang, L.; Chen, Y.; Yin, Y. Experimental and theoretical analysis of polymerization reaction process on the polydopamine membranes and its corrosion protection properties for 304 Stainless Steel. *J. Mol. Struct.* **2010**, *982*, 152–161. [CrossRef]
30. Luo, H.; Dong, C.F.; Li, X.G.; Xiao, K. The electrochemical behaviour of 2205 duplex stainless steel in alkaline solutions with different ph in the presence of chloride. *Electrochim. Acta* **2012**, *64*, 211–220. [CrossRef]
31. Lu, S.-Y.; Yao, K.-F.; Chen, Y.-B.; Wang, M.-H.; Ge, X.-Y. Influence of heat treatment on the microstructure and corrosion resistance of 13 wt PCT CR-type martensitic stainless steel. *Metall. Mater. Trans. A* **2015**, *46*, 6090–6102. [CrossRef]
32. Westin, E.M.; Olsson, C.-O.A.; Hertzman, S. Weld Oxide Formation on lean duplex stainless steel. *Corros. Sci.* **2008**, *50*, 2620–2634. [CrossRef]
33. Kokalj, A.; Lozinšek, M.; Kapun, B.; Taheri, P.; Neupane, S.; Losada-Pérez, P.; Xie, C.; Stavber, S.; Crespo, D.; Renner, F.U.; et al. Simplistic Correlations between Molecular Electronic Properties and Inhibition Efficiencies: Do They Really Exist? *Corros. Sci.* **2021**, *179*, 108856. [CrossRef]

Disclaimer/Publisher's Note: The statements, opinions and data contained in all publications are solely those of the individual author(s) and contributor(s) and not of MDPI and/or the editor(s). MDPI and/or the editor(s) disclaim responsibility for any injury to people or property resulting from any ideas, methods, instructions or products referred to in the content.

coatings

MDPI

Article

Morphological and Corrosion Characterization of Electroless Ni-P Coatings Deposited on Ductile Iron

Nicolás Ortiz [1], Jesús Rafael González-Parra [2,*], Jairo Olaya [1], Dayi Agredo [1], Raul Valdez [2], Helgi Waage [2], Ana María Bolarín [3], Félix Sánchez [3] and Arturo Barba-Pingarrón [2]

[1] Departamento de Ingeniería Mecánica y Mecatrónica—Bogotá, Facultad de Ingeniería, Universidad Nacional de Colombia, Bogota 16486, Colombia
[2] Centro de Ingeniería de Superficies y Acabados (CENISA), Facultad de Ingeniería, Universidad Nacional Autónoma de México, Mexico City 04510, Mexico; arturo.barba@ingenieria.unam.edu (A.B.-P.)
[3] Área Académica de Ciencias de la Tierra y Materiales, Universidad Autónoma del Estado de Hidalgo, Mineral de la Reforma 42184, Hidalgo, Mexico
* Correspondence: rparra@fisica.unam.mx

Abstract: Ductile iron is distinguished by its balance of mechanical properties and other advantageous characteristics, including its capacity to absorb energy. This makes it suitable for applications requiring high strength. However, its performance is impaired by its low corrosion resistance. In this study, a Ni-P coating was applied to the surface of ductile cast iron using electroless nickel plating to increase its corrosion resistance in 0.1 M NaCl. The characterization of the substrate and the coated materials was conducted using scanning electron microscopy, X-ray diffraction, electrochemical impedance spectroscopy, and electrochemical noise. The results show that the coating deposited on a ductile iron is amorphous, compact, homogeneous, and well-adhered. The surface hardness is increased by 53%. Also, a notable increase in corrosion resistance is evidenced by the blocking effect of the coating that delimits the access of the electrolyte to the ductile iron coating. The corrosion mechanism is related to the mixed and localized corrosion phenomena for the different evaluation times.

Keywords: ductile iron; electroless; coating; corrosion; electrochemical impedance spectroscopy

1. Introduction

Ductile or nodular cast iron is a material widely used in the automotive, textile, agricultural, and energy industries, among others. These Fe-C-Si metal alloys provide engineering parts with an attractive balance of mechanical properties such as tensile strength, toughness, mechanical vibration damping, good fluidity, good machinability, and good wear resistance [1,2]. The mechanical properties of ductile iron come from its chemical composition and microstructure, leading to a classification of five types, according to the mechanical properties and microstructure [3]. The typical microstructure consists of a matrix that can be ferritic, pearlitic, or ferritic–pearlitic, combined with spherical graphite nodules embedded in a ferrite zone. Ductile iron boasts better mechanical properties than grey cast iron due to the nodular form of the graphite, allowing for better continuity of the matrix, increasing the mechanical performance [4].

Ductile iron offers certain advantages due to its low cost and suitable mechanical properties. However, a significant number of the components manufactured with this material are adversely affected by the corrosive effects of the environment. When the integrity of the components is compromised, the functionality and operation of the machinery to which they belong is also affected. Consequently, corrosion resistance represents a significant challenge in the manufacturing of ductile iron engineering parts, given the considerable reduction in component lifespan that it entails [5,6].

The impact of thermal treatments and coating deposition through various techniques has been documented by numerous researchers, with notable enhancements in corrosion resistance observed. Some of these techniques include hot-dip galvanizing, thermal spraying, and the application of metallic coatings. The research conducted by Akinribide et al. [7] demonstrated that austempering heat treatment applied to nodular cast iron results in a reduction in the corrosion rate due to the stabilization of the ausferrite microstructure. Additionally, the research conducted by Pooja et al. [8] indicated that corrosion resistance was enhanced by austempering heat treatment. Specifically, the authors discovered that the corrosion resistance of the nodular cast iron was enhanced by increasing the temperature and austempering time due to the composition and microstructure of the material. Additionally, recent research has been conducted on the fabrication of coatings through the hot-dip galvanizing technique. Wu et al. [9] enhanced the coating thickness uniformity through the implementation of a selective pre-oxidation treatment of Si in nodular cast iron prior to coating. The authors found that the corrosion current density decreased significantly with an increasing selective treatment time, reaching a maximum polarization resistance value of 618.3 $\Omega{\cdot}cm^2$ after 5 h. In the research reported by Agredo et al. [10], the impact of a Zn-Al coating, obtained through thermal flame spraying, was examined on nodular cast iron. The authors indicate that the behavior of the coated material is more cathodic in comparison to the ductile iron substrate, resulting in the attainment of maximum corrosion potential values of 670 mV after 300 h.

On the other hand, electroless nickel plating comprises a series of techniques that allow for the deposition of coatings with uniform thicknesses on parts with irregular geometries which are difficult to access by other techniques. This type of coating is of significant interest due to its ability to be applied on a diverse range of materials, imparting exceptional corrosion and wear resistance to the substrate [11,12]. The electroless nickel plating technique offers a set of advantages such as simplicity, low cost, optimized temperatures, and suitable time parameters. All of these render this technique one of the most prominent methods for coatings technology, as well as one of the preferred methods for increasing the lifespan of various mechanical components. The characteristics of the electroless coating are modified by the P content and by the option to add other particles (Al_2O_3, SiC, TiO_2, SiO_2, CNT, TiC) that increase the wear and corrosion resistance of the coating due to the addition of nanoparticles that modify the morphology of the coating [13].

This process is carried out at low temperatures, representing a significant advantage over other techniques, as it ensures that the properties of the substrate remain unaltered by thermal effects. In the corrosion field, this technique has been applied in the electronics, automotive, oil and gas, aerospace, and food industries. This is due to the high corrosion protection that can be reached with the electroless nickel plating coating and its high hardness, which decreases the erosion of the surface [14]. Huang et al. [15] have reported the corrosion performance of mild steel with electroless nickel coating, indicating that this coating notably increases the corrosion resistance of the sample exposed to chloride and acidic media. Similar results have been reported by Seifzadeh et al. [16], which indicate the increase in corrosion resistance in a Ni-P coating with the co-deposition of cobalt. The addition of nanoparticles has been shown to cause a notable increment in the corrosion performance of this coating, as evidenced by Mazaheri et al. [17], who added nanometer-sized diamond to the Ni-P coating, increasing its corrosion resistance due to the changes in the coating morphology. On the other hand, another alternative for increasing the corrosion resistance is through the application of thermal treatments. Farzaneh et al. [18] have applied an annealing at 400 °C and 700 °C in a Ni-P coating, resulting in a higher corrosion resistance (41 $k\Omega cm^2$) for the sample annealed at 400 °C. In ductile iron, Li et al. [19] utilized a multilayer coating, applying the electroless nickel coating as an interlayer, with physical vapor deposition coatings. They indicated that this layer increased the corrosion resistance in NaCl by at least ten times, and that this interlayer can increase the adhesion strength between TiAlN/ZrN and the substrate. Thus, the application of this type of coating increases the corrosion resistance of several materials (metals, polymers, fabrics) and is a technique with industrial escalation potential. Despite the wide field of research regarding electroless nickel plating, there is a lack of information about the kinetics of the corrosion of these coatings in a popular ductile iron. In this study, the effect on the corrosion resistance of ductile iron after the generation of a Ni-P coating by electroless nickel plating was reported. This coating was evaluated by monitoring the corrosion behavior using surface, mechanical, and electrochemical non-destructive characterization methods.

2. Materials and Methods

The nodular cast iron was produced in the Metallurgical Engineering Department at the Universidad Nacional Autónoma de México and was kindly provided for the purposes of this study. To obtain the samples, a cast ingot was cut into cylindrical sections with a diameter of 2.5 cm and a thickness of 1 cm. The surface was ground in accordance with standard metallographic procedures (ASTM E3-11) using SiC paper with grain sizes of 80, 120, 240, 600, 1000, and 1500, polished with 1 μm alumina powder, and subsequently chemically etched with 3% Nital [20]. The elemental composition of the substrates was evaluated through the use optical emission spectrometry, employing BAIRD DV4 equipment.

The synthesis of the coatings was carried out following the methodology of Agredo et al. [12]. Briefly, an electroless chemical bath was prepared in distilled water, with the composition of nickel sulfate (0.2 M), sodium hypophosphite (0.25 M), propionic acid (0.1 M), lactic acid (0.25 M), sodium acetate (0.3 M), and metallic lead (1.3×10^{-6} M). The temperature was maintained at 88 ± 3 °C and pH 4.4 for 3.5 h, with continuous magnetic stirring throughout the process. Prior to immersion in the electroless bath, the surfaces were activated by immersing the samples in a 15% HCl distilled water solution for 5 min.

A microstructural study of the substrates and coatings was conducted using scanning electron microscopy (SEM) and energy dispersive spectroscopy (EDS). For this purpose, the samples were embedded in epoxy resin, and the cross section was polished, following the metallographic procedure described above. Carbon tape was employed to conduct the electrons and avoid the charge of the sample. The SEM equipment used was a Philips XL20 microscope (Eindhoven, The Netherlands), with an acceleration voltage of 15 kV. A semi-quantitative compositional analysis was applied to the coatings using an EDS microprobe coupled to an electron microscope. The morphology and EDS was measured using JEOL PC SEM v.5.1 software and Aztec 2.2 SP2, respectively. The crystalline phases of the ductile iron and the coatings produced were characterized by X-ray diffraction using a PANalytical X'PERT PRO MPD diffractometer (Malvern, UK) configured with Bragg–Brentano geometry. The step time was 30 s, the sweep range was 15–90°, and the scan time was 7 min at room temperature. The characterization parameters were obtained following the technique used in Ref. [21]. The mechanical properties of the substrate and the coating were evaluated for Vickers microhardness according to ASTM E384 [22] in an HVS-1000 microhardness tester, with a load of 100 g and a time of 10 s.

The electrochemical corrosion behavior of the system was assessed with an ACM Gill potentiostat/galvanostat using Sequencer version 5 software. After the coating was applied, the samples were washed and dried under a hot-air stream to clean the sample surface. The electrochemical cell configuration was the classic three-electrode arrangement, with a saturated calomel electrode (SCE) used as the reference electrode (RE), a graphite sheet as the counter electrode (CE), and the sample as the working electrode (WE). The samples were immersed in 0.1 M NaCl solution under static conditions for 504 h with an exposed area of 2.54 cm^2. For the corrosion testing, the exposed area of the WE was controlled using an O-ring, and the sample was electrically connected to a Caiman clamp to provide electrical contact. For electrochemical impedance spectroscopy (EIS), the frequency was swept between 10^4 Hz and 10^{-1} Hz using a sinusoidal signal with amplitude of 10 mV rms, acquiring 10 data points per decade. For electrochemical noise (EN), time series of 1024 points at 0.5 s intervals were obtained. The DC trend of the time series was eliminated using the linear method [23]. The noise resistance parameter (Rn) [24] and the localization index [25], which identifies the corrosion mechanism, were computed.

3. Results and Discussion

3.1. Substrate and Coating Characterization

Table 1 shows the composition determined for the metallic substrates. The elemental study of the casting demonstrates a predominant composition of Fe-C-Si. Additionally, several components were identified. The primary elements, Fe-C-Si, are indicative of this specific casting type and determine the properties of the material [10].

Table 1. Elemental composition of cast iron (% weight).

C	Si	Mn	P	S	Cr	Mo	Ni	Al	Co	Cu	Nb	Ti	V	Pb	Mg	Fe
3.8	2.5	0.3	0.01	0.01	0.08	<0.05	0.03	0.01	0.005	0.14	0.003	0.003	0.005	0.002	0.03	Bal.

The microstructure of the substrate (Figure 1) shows the presence of a pearlitic matrix consisting of alternating layers of ferrite and cementite phases. Additionally, a soft ferrite phase is observed around the graphite nodules. This microstructure corresponds to the typical microstructure of a ductile iron. In this type of material, pearlite is a hard structure, and ferrite is a soft structure, providing an optimal combination of strength and toughness properties. These properties are also influenced by the presence of nodular graphite [26,27].

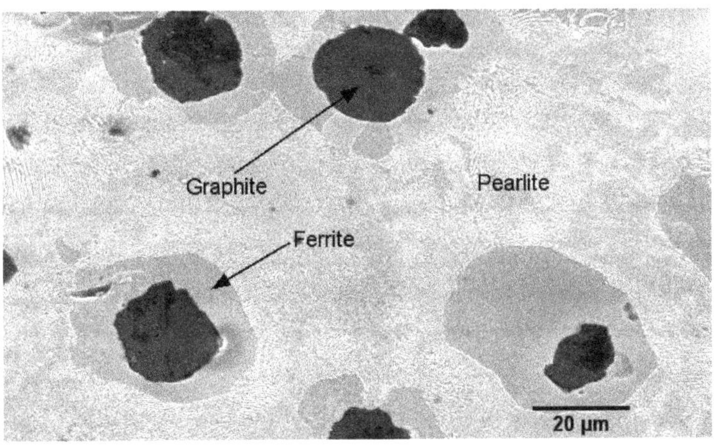

Figure 1. Nodular cast iron microstructure. SEM image with secondary electrons (SE) 800×.

Figure 2a shows the cross section of the metallic samples. The coating shows a well-defined layer adhered over the whole substrate area. This behavior is related to the strong chemical bonding between the coating and the substrate due to the chemical reduction reactions carried out during the coating formation. There is no evidence of either porosity or delamination, indicating the high homogeneity of the coating [28]. The average coating thickness was 34.89 ± 0.36 µm. The EDS results (Figure 2b) show the presence of characteristic peaks for Ni and P and a weight percent of 83.9% and 16% for Ni and P, respectively. These results confirm the formation of a Ni-P alloy and are in agreement with the composition results obtained by Zeng et al. [29]. Forestier et al. [30] applied an electroless nickel plating to EN-GJS-500-7 ductile iron over a period of 60 min, reporting a localized decrease in thickness around the graphite nodules; however, in our case, this behavior was not evidenced, with the plating reaching a homogeneous thickness after 210 min of coating formation.

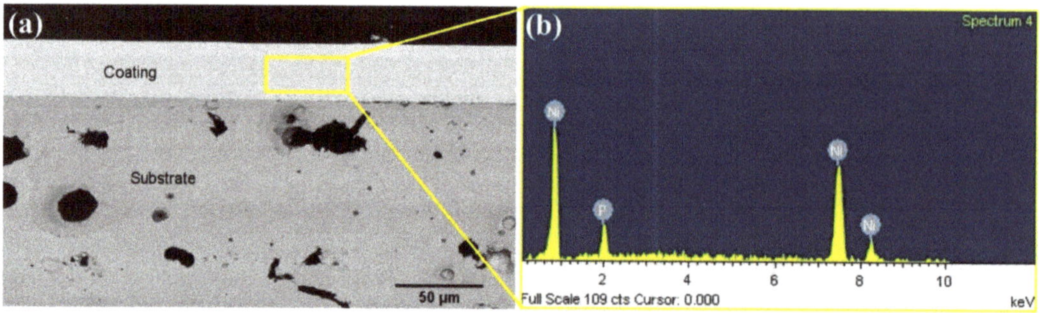

Figure 2. (**a**) The coating on ductile iron and the EDS spectra; (**b**) SEM image with backscattered electrons (BSE) 400×.

The mechanical properties of the substrate and the coating were evaluated using Vickers microhardness, as shown in Table 2. The overall material hardness was 311.99 ± 1.19 HV. Based on these results and the observed microstructure, it can be concluded that the casting corresponds to a nodular cast iron grade of 100-70-3 (ASTM A 536), associated with a material tensile strength of 690 MPa [3]. On the other hand, the coating has a value of 467.66 ± 29.63 HV, which represents an increase in hardness of 53.28% with respect to the substrate. This represents an improvement in terms of the wear resistance of the base material, and these results are in agreement with those reported in Ref. [31].

Table 2. Vickers microhardness of the samples.

Phases	Hardness (HV)
Pearlite	380.48 ± 10.38
Ferrite	170.02 ± 4.24
Graphite	154.04 ± 1.25
Substrate (Global)	311.99 ± 1.19
Ni-P coating	467.66 ± 29.63

X-ray diffraction was used to characterize the structure of the samples. Figure 3a shows the diffraction pattern of the substrate. The plot shows the characteristic peaks associated with the crystalline structure of Fe and C (graphite), with semi-quantitative phase percentages of 82 and 18%, respectively. The Fe has a cubic crystalline structure within the Im-3m space group, with two diffraction maxima located at 52.49 and 77.49°. This result is in agreement with the data from the ICSD 01-087-0722 reference. The diffraction pattern of the carbon (graphite) sample shows a single peak at $2\theta = 31.01°$, with a rhombohedral crystalline structure and an R-3m space group, as indicated by the ICSD 01-075-2078 reference pattern. Thus, the substrate is composed of Fe related to the ferrite phase (body-centered cubic) and graphite, indicating the composition of the ductile iron. In contrast, Figure 3b shows the diffraction pattern of the Ni-P coating, which displays a structure of predominantly amorphous nature, with a ratio of 80% amorphous and 20% crystalline structure. The peak of this structure is located at $2\theta = 44.51°$, corresponding to the reference pattern ICSD 00-004-0850, and coincides with a Ni peak of a face-centered cubic (FCC) structure in the (111) plane. The formation of an amorphous structure can be attributed to the lattice distortion experienced by the nickel crystal structure due to the presence of phosphorus atoms. This structure hinders the propagation of the cubic occupation centered on the faces of the nickel atoms [31,32], which is supported by the high phosphorus content of the coating. At levels higher than 10%, the structure of the material is conditioned to an amorphous structure [33]. The formation of an amorphous coating is beneficial to the corrosion resistance of the samples, since it possesses an absence of grain boundaries [34].

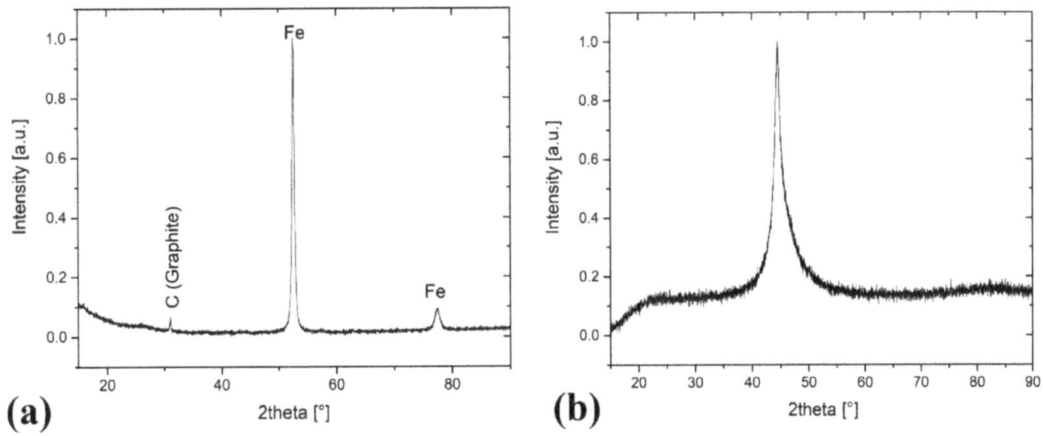

Figure 3. Diffraction pattern: (**a**) nodular cast iron; (**b**) Ni-P coating.

3.2. Corrosion Analysis

To assess the corrosion behavior, the samples were continuously immersed in a 0.1 M NaCl solution for 504 h. Figure 4 show the EIS response of the samples throughout the corrosion process. The Nyquist plot, Figure 4a,b, and the Bode plots, Figure 4c,d, are illustrative representations that facilitate the interpretation of the data. The Nyquist plot depicts the real component (Z′) versus the imaginary component (-Z″) of the impedance for the substrates and coatings. In this plot, the two systems (the blank and the Ni-P coatings) exhibit a resistive–capacitive semi-arc with varying diameters that evolve with the immersion time. As observed, the Nyquist plot for the substrate (Figure 4a) shows a reduction in the diameter of the semi-arc as corrosion time increases. In the field of corrosion literature, there is a commonly accepted understanding that the point of intersection of the semi-arc with the real axis represents the overall impedance of the system, thereby facilitating the association with corrosion resistance [35].

Consequently, as the corrosion time increases, the corrosion resistance of the substrate is diminished. In contrast, the coated material (Figure 4b) show a notable increase in the semi-arc diameter with increasing immersion time. As the immersion time increases up to 168 h, the corrosion resistance reaches a maximum, after which it undergoes a slight decrease [20]. It is noteworthy that the addition of the coating increases the impedance values of the system, which suggests an improvement in corrosion performance. This behavior is clearly evidenced in the impedance modulus Bode plot representation, shown in Figure 4c. The impedance modulus exhibits two distinct groups. The first is within the $|Z|$ range of 10^2 to 10^3 ohm cm^2, while the second is nearly within the range of 10^3 to 10^4 Hz. This suggests that the uncoated and coated samples show notable differences in impedance values, obtaining higher impedance values for the coated samples by at least one order of magnitude. The low-frequency values are associated with the polarization resistance of the system, indicating that the coated samples exhibit enhanced corrosion resistance. These results are in concordance with the behavior of the Nyquist plots.

Figure 4d illustrates the phase angle plots. The results for the substrate indicate the appearance of a localized time constant at low frequencies, with a maximum observed around 10^{-1} to 10^0 Hz. In contrast, the coated material shows the appearance of a localized time constant at medium frequencies, which shifts towards the low-frequency region. In the case of the coated sample, an incipient and flat time constant related to the coating was considered. The shift of the maximum towards lower frequencies is indicative of a reduction in the corrosion phenomenon. Therefore, it can be understood that the addition of the coating has a mitigating effect on corrosion. Conversely, there is no evidence of the formation of a protective corrosion products layer in the substrate, indicating that corrosion proceeds unimpeded. These results are in concordance with the Nyquist and impedance modulus plots, which do not indicate an increase in the total impedance of the system. Additionally, the coated material exhibits the formation of a protective layer of corrosion products. According to Pan et al. [36], these products were formed by the interaction of Fe^{2+} and Fe^{3+} ions with PO_4^{3-}, which promoted the passivation of the surface. Moreover, the EIS results were fitted to the equivalent electrical circuits, as shown in Figure 5a,b. Based on the results obtained, the circuit in Figure 5a was used for adjusting the uncoated specimens, while the circuit in Figure 5b was used for the coated specimen after 168 h of immersion. The parameters used are described as follows: Rs is the solution resistance, Rct represents the charge transfer resistance, CPEdl is the constant phase element related to the double layer, Rf is the resistance of the formed film, and CPEf is the constant phase element associated with the formed corrosion product layer. The circuits used were reported by Li et al. [37] for their study of coatings similar to those used in this work.

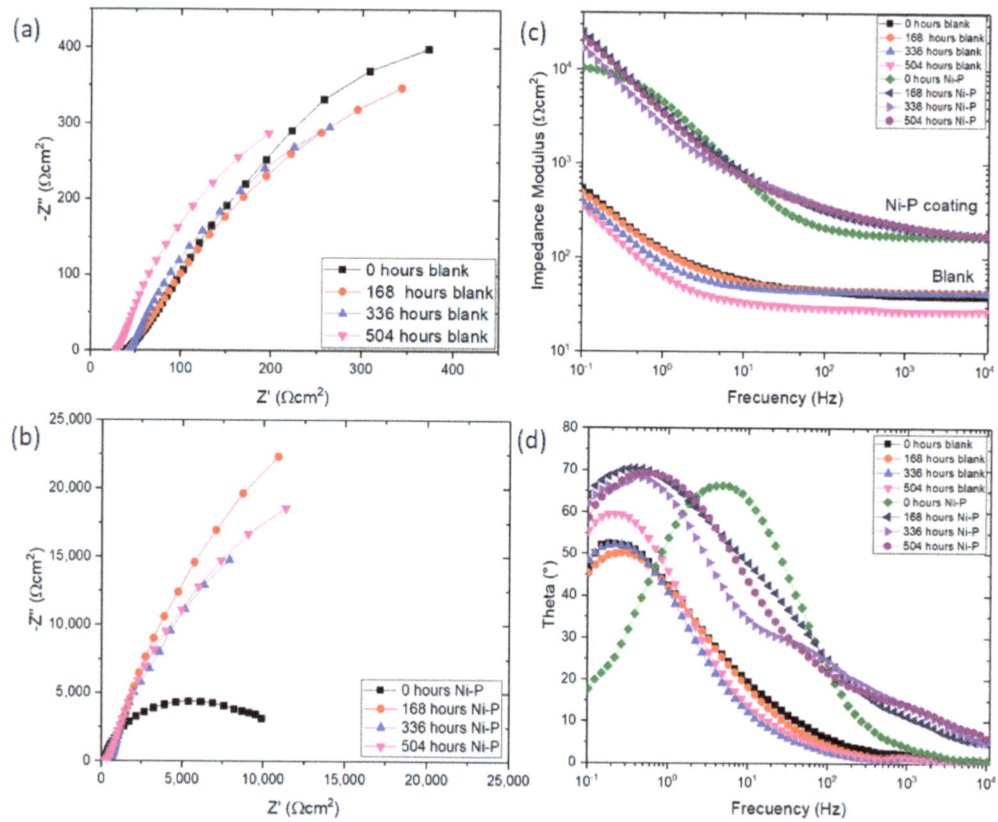

Figure 4. Nyquist diagram: (**a**) substrate; (**b**) Ni-P coating; (**c**) impedance modulus; (**d**) phase angle.

As part of the adjustment, a constant phase element (CPE) was used to adapt the capacitive behavior, since the CPE shows sensitivity to surface conditions, such as roughness. The impedance of the CPE is represented by the following equation:

$$Z_{CPE} = \left[Y_{CPE}(jw)^{n}\right]^{-1} \quad (1)$$

where Y represents the admittance of CPE, w represents the angular frequency, and n represents the coefficient associated with the capacitance deviation. This coefficient can have values from 0 to 1, indicating fully resistive or capacitive behavior, respectively [38].

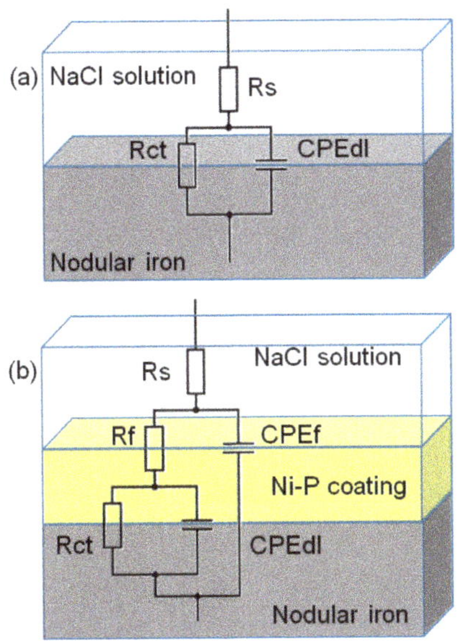

Figure 5. Equivalent electrical circuits: (**a**) substrate; (**b**) Ni-P coating.

The results of the fitting are presented in Table 3. Additionally, the final column specifies the electric equivalent circuit utilized for the fitting results. The results obtained for the uncoated material show an increase in the corrosion system. This behavior is evidenced by the reduction in charge transfer resistance, as well as by the behavior of the CPE and the parameter n. This is associated with the decrease in the corrosion resistance of the system [39]. In contrast, for the coated material, the Rf and CPEf parameters show the formation of a layer of corrosion products after 168 h [40]. However, the Rf behavior associated with the dissolution of the corrosion products formed is also evidenced by the behavior of Rct, which reaches its maximum at 336 h of immersion and then decreases. This behavior agrees with that previously observed in the Nyquist diagram.

Table 3. Parameters obtained from the EIS adjustment and electric equivalent circuits (EEC) employed.

	Time	CPEf	nf	Rf	CPE dl	ndl	Rct	Chi Squared	EEC
	h	Mohs·S^n		Ω·cm^2	Mohs·S^n		kΩ·cm^2	(×10^{-3})	
Substrate	0	--	--	--	2.8×10^{-3}	0.66	8.3	9.8	a
	168	--	--	--	2.8×10^{-3}	0.71	1.9	3.2	a
	336	--	--	--	3.5×10^{-3}	0.79	1.2	5.4	a
	504	--	--	--	4.4×10^{-3}	0.80	1.8	14.9	a
Ni-P Coating	0				3.8×10^{-5}	0.89	10.8	19.5	a
	168	2.5×10^{-5}	0.75	298.8	3.7×10^{-5}	0.80	150	2.5	b
	336	5.5×10^{-5}	0.66	893.5	3.7×10^{-5}	0.98	135	3.2	b
	504	2.2×10^{-5}	0.73	321	4.2×10^{-5}	0.86	106.6	3.1	b

In summary, the polarization resistance ($R_{pol} = R_f + R_{ct}$) was calculated based on the fitted parameters shown in Table 3. Figure 6 show the polarization resistance of the coated and uncoated samples immersed for 500 h in a chloride-containing corrosive solution. The Ni-P coating samples show an R_{pol} of almost two orders of magnitude higher than that of the blank sample. Consequently, the coating increases the total resistance of the system due to the synergistic effect of the increased film resistance and the detriment from the charge transfer.

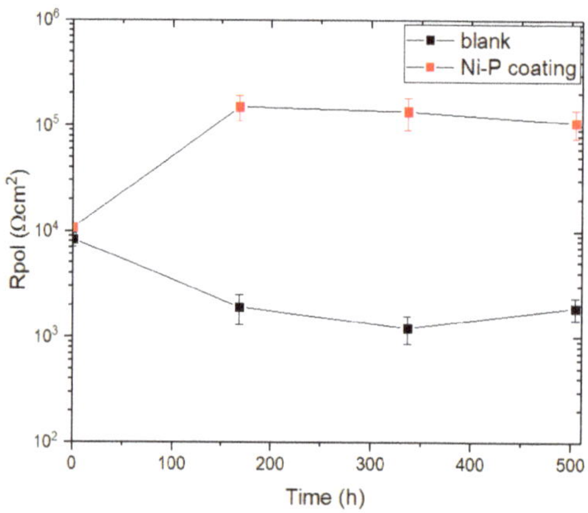

Figure 6. Polarization resistance of the samples immersed in NaCl solution.

Figure 7a shows the results of the noise resistance; this parameter was calculated using the following equation $R_n = {}^{\sigma_E}/_{\sigma_I}$, where σ_E and σ_I represent the potential standard deviation and current standard deviation, respectively. The substrate shows a decreasing behavior as a function of the immersion time. These results are consistent with those of the polarization resistance obtained through EIS. As the immersion time increases, the samples exhibit degradation due to the influence of the electrolyte. The coating on the cast iron shows a trend of increasing noise resistance up to 168 h of evaluation. After this period, degradation occurs, yet the R_n remains higher in comparison with the initial evaluation and is significantly above the R_n of the substrate. As evidenced in the literature, noise resistance is clearly correlated with corrosion resistance [41]. Then, this behavior suggests that the addition of the electroless Ni-P coating facilitates the formation of a protective corrosion products layer, thereby enhancing the corrosion resistance of the samples.

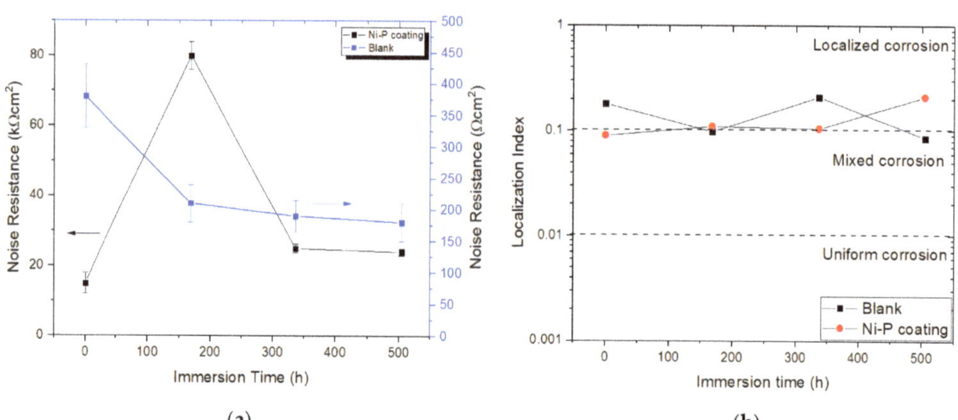

Figure 7. (**a**) Noise resistance (Rn) and (**b**) localization index of the samples immersed in NaCl solution.

In Figure 7b, the substrate shows a localization index that varies between the localized corrosion for 0 h and 336 h and mixed corrosion for 168 h and 504 h. These corrosion phenomena make the material more susceptible to degradation. The results also indicate that the layer of corrosion products has no protective effect on the blank material. In comparison, the coating begins with a mixed corrosion process that change to a localized corrosion process. At 336 h, the corrosion mechanism of the coated metal falls between mixed and localized corrosion, and at 504 h, the behavior changes completely to localized corrosion. The behavior of this system is due to the initial appearance of a protective layer, followed by fracture and pitting. Although the coating is highly protective and regenerates, it does not prevent the corrosion process from continuing [42].

4. Conclusions

The coating generated exhibits high compactness, with good homogeneity in its thickness, thus presenting an average value of approximately 35 μm and hardness on the order of 468 HV.

The EDS and XRD studies allow for the determination of the formation of a Ni-P coating of predominantly amorphous nature. This phenomenon, linked to the generation of corrosion products, allows the coated material to exhibit a substantial improvement in corrosion resistance, as is evident in the case that reached the point of greatest resistance at 168 h of immersion with a value of Rn = 116.59 k$\Omega \cdot cm^2$.

The substrate shows a combined corrosion phenomenon. First, a mixed corrosion phenomenon is observed at 168 h and 504 h, followed by a localized corrosion phenomenon at 0 h and 336 h of immersion. On the other hand, the coating exhibits only an initial mixed corrosion phenomenon at 0 h of evaluation, which transforms into a localized corrosion process over the subsequent hours of immersion.

The corrosion products generated in the ductile iron turned out to be porous and not very adherent, which allowed the corrosion process to continue. The coatings exhibited a behavior of increased corrosion resistance, evidenced by the increase in the polarization resistance, which was linked to the generation of a compact and highly adherent layer of corrosion products.

Author Contributions: Data curation, R.V., H.W. and F.S.; formal analysis, J.R.G.-P.; funding acquisition, A.B.-P.; investigation, N.O. and D.A.; methodology, D.A.; project administration, J.O. and A.B.-P.; resources, J.O., A.M.B. and A.B.-P.; validation, F.S.; visualization, H.W.; writing—original draft, N.O. and D.A.; writing—review and editing, J.R.G.-P. All authors have read and agreed to the published version of the manuscript.

Funding: This research received no external funding.

Institutional Review Board Statement: Not applicable.

Informed Consent Statement: Not applicable.

Data Availability Statement: Data are contained within the article.

Acknowledgments: The authors thank the Department of Mechanics and Mechatronic Engineering of Universidad Nacional de Colombia—Sede Bogotá for providing access to their laboratories, as well as research support. Special thanks to Adriana Tejeda Cruz from IIM-UNAM, X-ray Laboratory, for her valuable support in the structural characterization of the coatings. Also, the authors thanks to the support of the Dirección General de Asuntos del Personal Académico of UNAM, through the PAPIIT IT101221 project "Desarrollo de Tecnologías Alternativas de Modificación Superficial para la Mejora de Materiales de Potencial Industrial".

Conflicts of Interest: The authors declare no conflict of interest.

References

1. Chikali, P.B.; Shinde, V.D. Analysis of machinability in ductile iron casting. *Mater. Today Proc.* **2019**, *27*, 584–588. [CrossRef]
2. Labrecque, C.; Gagné, M. Ductile Iron: Fifty years of continuous development. *Can. Metall. Q.* **1998**, *37*, 343–378. [CrossRef]
3. *ASTM A536*; Standard Specification for Ductile Iron Castings. ASTM International: West Conshohocken, PA, USA, 2009.

4. Concli, F. Austempered Ductile Iron (ADI) for gears: Contact bending Austempered Ductile Iron (ADI) for gears: Contact and bending fatigue behavior fatigue behavior Thermo-mechanical modeling of a high pressure turbine. *Procedia Struct. Integr.* **2018**, *8*, 14–23. [CrossRef]
5. Song, Y.; Jiang, G.; Chen, Y.; Zhao, P.; Tian, Y. Effects of chloride ions on corrosion of ductile iron and carbon steel in soil environments. *Sci. Rep.* **2017**, *7*, 6865. [CrossRef]
6. Wasim, M.; Djukic, M.B. Corrosion induced failure of the ductile iron pipes at micro- and nano-levels. *Eng. Fail. Anal.* **2021**, *121*, 105169. [CrossRef]
7. Akinribide, O.J.; Akinwamide, S.O.; Ajibola, O.O.; Obadele, B.A.; Olusunle, S.O.O.; Olubambi, P.A. Corrosion behavior of ductile and austempered ductile cast iron in 0.01M and 0.05M NaCl Environments. *Procedia Manuf.* **2019**, *30*, 167–172. [CrossRef]
8. Pooja, M.; Vijeesh, V.; Surendranathan, A.O.; Udupa, K.R.; Samuel, K.G. Corrosion behavior of austempered ductile iron (ADI) in iron ore slurry. *Int. J. Eng. Sci. Technol.* **2016**, *8*, 7. [CrossRef]
9. Wu, H.; Peng, H.; Liu, Y.; Tu, H.; Wang, J.; Su, X. Effect of Selective Pre-oxidation on the Growth and Corrosion Resistance of Hot-Dip Zinc Coating of Nodular Cast Iron. *Corros. Sci.* **2021**, *186*, 109463. [CrossRef]
10. Diaz, D.G.A.; Godoy, N.O.; Navarro, R.G.V.; Pingarrón, A.B.; Flórez, J.J.O.; Parra, J.R.G.; Villar, A.C.; Gallegos, M.Á.H. Caracterización electroquímica de recubrimientos Zn-Al sobre fundición nodular grado 2, obtenidos por proyección térmica por flama con alambre. *Av. Investig. En Ing.* **2020**, *17*, 5747. [CrossRef]
11. Narayanan, T.S.N.S.; Baskaran, I.; Krishnaveni, K.; Parthiban, S. Deposition of electroless Ni-P graded coatings and evaluation of their corrosion resistance. *Surf. Coatings Technol.* **2006**, *200*, 3438–3445. [CrossRef]
12. Diaz, D.G.A.; Pingarrón, A.B.; Florez, J.J.O.; Parra, J.R.G.; Cabello, J.C.; Moncaleano, I.A.; Villar, A.C.; Gallegos, M.Á.H. Effect of a Ni-P coating on the corrosion resistance of an additive manufacturing carbon steel immersed in a 0.1 M NaCl solution. *Mater. Lett.* **2020**, *275*, 128159. [CrossRef]
13. Nazari, H.; Darband, G.; Arefinia, R. A review on electroless Ni-P nanocomposite coatings: Effect of hard, soft, and synergistic nanoparticles. *J. Mater. Sci.* **2023**, *58*, 4292–4358. [CrossRef]
14. Chintada, V.; Koona, R.; Bahubalendruni, M. State of art review on Nickel-based electroless coatings and materials. *J. Bio Tribo-Corros.* **2021**, *7*, 134. [CrossRef]
15. Huang, Y.; Zeng, X.; Hu, X.; Liu, F. Corrosion resistance properties of electroless nickel composite coatings. *Electrochim. Acta* **2004**, *49*, 4313–4319. [CrossRef]
16. Seifzadeh, D.; Hollagh, A. Corrosion resistance enhancement of AZ91D magnesium alloy by electroless Ni-Co-P coating and Ni-Co-PSiO2 nanocomposite. *J. Mater. Eng. Perform.* **2014**, *23*, 4109–4121. [CrossRef]
17. Mazaheri, H.; Allahkaram, S. Deposition, characterization and electrochemical evaluation of Ni-P-nano diamond composite coatings. *Appl. Surf. Sci.* **2012**, *258*, 4574–4580. [CrossRef]
18. Farzaneh, A.; Ehteshamzadeh, M.; Can, M.; Mermer, O.; Okur, S. Effects of SiC particles size on electrochemical properties of electroless Ni-P-SiC nanocomposite coatings. *Prot. Met. Phys. Chem. Surf.* **2016**, *52*, 632–636. [CrossRef]
19. Lin, C.; Hsu, C.; Cheng, Y.; Ou, K.; Lee, S. A study on the corrosion and erosion behavior of electroless nickel and TiAlN/ZrN duplex coatings on ductile iron. *Appl. Surf. Sci.* **2015**, *324*, 13–19. [CrossRef]
20. *ASTM E3-11*; Preparation of Metallographic Specimens. ASTM International: West Conshohocken, PA, USA, 2017.
21. Diaz, D.G.A.; Pingarrón, A.B.; Florez, J.J.O.; Parra, J.R.G.; Cabello, J.C.; Moncaleano, I.A.; Villar, A.C.; Gallegos, M.Á.H. Evaluation of the corrosion resistance of a Ni-P coating deposited on additive manufacturing steel: A dataset. *Data Br.* **2020**, *32*, 106159. [CrossRef]
22. *ASTM E384*; Standard Test Methods for Microindentation Hardness of Materials. ASTM International: West Conshohocken, PA, USA, 1999.
23. Lentka, L.; Smulko, J. Methods of trend removal in electrochemical noise data—Overview. *Meas. J. Int. Meas. Confed.* **2019**, *131*, 569–581. [CrossRef]
24. *G199-09*; Standard Guide for Electrochemical Noise Measurement. ASTM International: West Conshohocken, PA, USA, 2014.
25. Xia, D.H.; Song, S.Z.; Behnamian, Y. Detection of corrosion degradation using electrochemical noise (EN): Review of signal processing methods for identifying corrosion forms. *Corros. Eng. Sci. Technol.* **2016**, *51*, 527–544. [CrossRef]
26. Wang, B.; Qiu, F.; Barber, G.C.; Pan, Y.; Cui, W.; Wang, R. Microstructure, wear behavior and surface hardening of austempered ductile iron. *J. Mater. Res. Technol.* **2020**, *9*, 9838–9855. [CrossRef]
27. Olawale, J.O.; Ibitoye, S.A. Processing Techniques and Productions of Ductile Iron: A Review. *Int. J. Sci. Eng. Res.* **2016**, *7*, 397–423.
28. Chang, C.S.; Hou, K.H.; Der Ger, M.; Chung, C.K.; Lin, J.F. Effects of annealing temperature on microstructure, surface roughness, mechanical and tribological properties of Ni-and Ni-P/SiC films. *Surf. Coat. Technol.* **2016**, *288*, 135–143. [CrossRef]
29. Zeng, L.; Wang, C.; Li, Y.; Hu, T. Synthesis of core-shell Ti@Ni-P spherical powder by Ni electroless plating. *Micron* **2021**, *143*, 103027. [CrossRef]
30. Forestier, I.; Berthomé, G.; Wouters, Y. Study of electroless Nickel coatings on EN-GJS-500-7 spheroidal graphite cast iron. *Coatings* **2018**, *8*, 239. [CrossRef]
31. Fayyaz, O.; Khan, A.; Shakoor, R.A.; Hasan, A.; Yusuf, M.M.; Montemor, M.F.; Rasul, S.; Khan, K.; Faruque, M.R.I.; Okonkwo, P.C. Enhancement of mechanical and corrosion resistance properties of electrodeposited Ni–P–TiC composite coatings. *Sci. Rep.* **2021**, *11*, 5327. [CrossRef]

32. Yang, Y.; Chen, W.; Zhou, C.; Xu, H.; Gao, W. Fabrication and characterization of electroless Ni–P–ZrO_2 nano-composite coatings. *Appl. Nanosci.* **2011**, *1*, 19–26. [CrossRef]
33. Gomes, N.; González-Estrada, O.A.; Pertuz, A. Electroless Nickel Phosphorous: Una visión global. *Rev. UIS Ing.* **2019**, *18*, 173–192. [CrossRef]
34. Fayyad, E.; Abdullah, A.; Hassan, M.; Mohamed, A.; Jarjoura, G.; Farhat, Z. Recent advances in electroless-plated Ni-P and its composites for erosion and corrosion applications: A review. *Emergent Mater.* **2018**, *1*, 3–24. [CrossRef]
35. Chang, B.Y.; Park, S.M. Electrochemical impedance spectroscopy. *Annu. Rev. Anal. Chem.* **2010**, *3*, 207–229. [CrossRef] [PubMed]
36. Pan, C.; Song, Y.; Jin, W.; Qin, Z.; Song, S.; Hu, W.; Xia, D.H. Enhancing the Stability of Passive Film on 304 SS by Chemical Modification in Alkaline Phosphate–Molybdate Solutions. *Trans. Tianjin Univ.* **2020**, *26*, 135–141. [CrossRef]
37. Li, J.; Sun, C.; Roostaei, M.; Mahmoudi, M.; Fattahpour, V.; Zeng, H.; Luo, J.L. Characterization and corrosion behavior of electroless Ni-Mo-P/Ni-P composite coating in CO_2/H_2S/Cl− brine: Effects of Mo addition and heat treatment. *Surf. Coatings Technol.* **2020**, *403*, 126416. [CrossRef]
38. Tafreshi, M.; Allahkaram, S.R.; Farhangi, H. Comparative study on structure, corrosion properties and tribological behavior of pure Zn and different Zn-Ni alloy coatings. *Mater. Chem. Phys.* **2016**, *183*, 263–272. [CrossRef]
39. Liao, X.N.; Cao, F.H.; Chen, A.N.; Liu, W.J.; Zhang, J.Q.; Cao, C.N. In-situ investigation of atmospheric corrosion behavior of bronze under thin electrolyte layers using electrochemical technique. *Trans. Nonferrous Met. Soc. China* **2012**, *22*, 1239–1249. [CrossRef]
40. Marušić, K.; Ćurković, H.O.; Takenouti, H. Inhibiting effect of 4-methyl-1-p-tolylimidazole to the corrosion of bronze patinated in sulphate medium. *Electrochim. Acta* **2011**, *56*, 7491–7502. [CrossRef]
41. Bethencourt, M.; Gonz, L.; Botana, F.J. Noise resistance and shot noise parameters on the study of IGC of aluminium alloys with different heat treatments. *Electrochim. Acta* **2007**, *52*, 6569–6583. [CrossRef]
42. Domínguez, A.S.; Bueno, J.J.P.; Torres, I.Z.; López, M.L.M. Corrosion study of electroless deposited nickel-phosphorus solar absorber coatings on carbon steel. *Int. J. Electrochem. Sci.* **2017**, *12*, 2987–3000. [CrossRef]

Disclaimer/Publisher's Note: The statements, opinions and data contained in all publications are solely those of the individual author(s) and contributor(s) and not of MDPI and/or the editor(s). MDPI and/or the editor(s) disclaim responsibility for any injury to people or property resulting from any ideas, methods, instructions or products referred to in the content.

coatings

MDPI

Article

Individual and Joint Effect of Oleic Acid Imidazoline and $CeCl_3$ on Carbon Steel Corrosion in CO_2-Saturated Brine Solution

Tihomir Borko [1], Gordana Bilić [2,*], Katarina Žbulj [2] and Helena Otmačić Ćurković [3]

[1] INA, d.d., 10000 Zagreb, Croatia; tihomir.borko@ina.hr
[2] Faculty of Mining, Geology and Petroleum Engineering, University of Zagreb, 10000 Zagreb, Croatia; katarina.zbulj@rgn.unizg.hr
[3] Faculty of Chemical Engineering and Technology, University of Zagreb, 10000 Zagreb, Croatia; hotmac@fkit.unizg.hr
* Correspondence: gordana.bilic@rgn.hr; Tel.: +385-1-5535-909

Abstract: In production and transportation systems of the oil industry, brine solutions contain high concentrations of chloride and dissolved CO_2, which is a very corrosive medium to which carbon steel is exposed. Therefore, finding new effective and environmentally friendly corrosion inhibitors is of great importance. The effect of $CeCl_3$ (in concentrations from 5 mg dm^{-3} to 20 mg dm^{-3}) and oleic acid imidazoline (IOA) (in concentrations from 5 mg dm^{-3} to 20 mg dm^{-3}) separately and their mixtures (in concentrations from 5 mg dm^{-3} to 15 mg dm^{-3} of $CeCl_3$ and from 5 mg dm^{-3} to 20 mg dm^{-3} of IOA) as corrosion inhibitors of AISI 1018 carbon steel corrosion in simulated brine solution saturated with CO_2 at 60 °C were examined by means of weight-loss testing, electrochemical measurements (polarization resistance, linear polarization with Tafel extrapolation, electrochemical impedance spectroscopy) and surface analyses (scanning electron microscopy with energy-dispersive X-ray spectroscopy analyses, Raman spectroscopy and X-ray diffraction). All test methods showed a higher efficiency of compounds' mixtures (from 62.77% to 97.94%) and a higher degree of corrosion protection compared to the action of individual compounds (efficiency from 3.43% to 94.61% for IOA and from 57.58% to 96.27% for $CeCl_3$). Imidazoline, a common corrosion inhibitor in CO_2-saturated systems, most likely forms a surface film with voids via its adsorption on steel surface, while cerium carbonate tends to fill these voids by creating a more compact film. In this way, a denser and thicker surface film is formed.

Keywords: CO_2 corrosion; corrosion inhibitor; $CeCl_3$; oleic acid-based imidazoline; synergism

1. Introduction

Carbon capture, utilization and storage (CCUS) is considered an effective way of reducing CO_2 emissions, and it is often used for enhanced oil recovery purposes (CO_2-EOR) [1–6]. It is known that the produced fluids are quite corrosive media (they contain oil, water, chlorides, acids and dissolved gases) that impair the integrity and useful properties of transportation pipelines and tubing [7–9].

In oil and gas production and transport pipelines, as well as in carbon capture storage systems, corrosion of the inner walls of the pipes, due to the presence of CO_2, results in pipeline rupture, the loss of production and transport fluid, economic losses and ecological disaster [10–14]. Situations are much worse when high chloride concentrations are present as well as turbulent hydrodynamic conditions [15].

Most pipelines for systems containing CO_2 (oil and gas production, geothermal systems, carbon capture, utilization and storage systems) are made of carbon steel. Although carbon steel is subject to corrosion in exposed conditions, the reason for its wide application is due to its favorable metallurgical properties as well as economic profitability [16–18].

One of the oldest and most common methods of protecting metals from corrosion is the application of corrosion inhibitors [19]. Corrosion inhibitors are defined as substances that, added in small amounts, slow down or prevent metal corrosion [20]. In the past, the only factor in selecting an inhibitor was its effectiveness, while its environmental impact was neglected. In addition, it was believed that effective corrosion inhibitors could not be environmentally friendly, and environmentally friendly compounds could not be effective corrosion inhibitors [21].

Fortunately, nowadays, the selection and application of corrosion inhibitors has moved from an exclusively economic aspect to taking into account environmental factors, i.e., it is necessary to satisfy the so-called E3 concept (efficiency, economy and ecology). Before 1960, the only criterion that had to be met when choosing an inhibitor was effectiveness, and the following were widely used: chromates, phosphates, nitrates, borates, silicates and zinc compounds. Although extremely effective as corrosion inhibitors and economically viable, today's chromates (like most inorganic inhibitors) are banned for health and safety reasons. From 1960 to 1980, the focus when choosing inhibitors was on the economic aspect; so polyphosphates, gluconates, vanadates, molybdates, carboxylates and surface-active chelates were used. Tannins, vitamins, natural compounds and polymers were widely used from 1980 to 1995 when ecology was the main required property of the inhibitor. From 1995 until today, inhibitors must be environmentally friendly, which is why compounds like rare earth metal compounds, multifunctional organic compounds, mixtures of rare earth metals and organic/inorganic compounds and the encapsulation of inhibitors are preferred [21,22].

Chemicals in the general sense, and thus non-ecological corrosion inhibitors, cause numerous environmental problems, which is why the development of "green chemistry" that also applies to corrosion inhibitors has occurred. "Green chemistry" refers to chemicals and chemical processes that reduce or eliminate the negative impact on the environment since they reduce waste products, do not contain toxic compounds and improve efficiency [23].

Scientists invest a lot of effort in finding and producing inhibitors that are both highly effective and low in toxicity. Such inhibitors are called "environmentally friendly", "green inhibitors" or "environmentally acceptable inhibitors" [24,25].

In the eighties of the last century, the development of a new group of corrosion inhibitors, salts of rare earth metals (REM), began. Their development was slow, and they did not reach the importance of organic inhibitors despite their significant effectiveness. Among the rare earth salts, lanthanoids stand out as effective corrosion inhibitors. Cerium, praseodymium, neodymium, lanthanum and ytterbium ions showed very high inhibitory effectiveness on aluminum alloys [26,27].

Small animal studies of cerium compounds have shown low toxicity, which can be related to other lanthanides. Lanthanides are economically competitive products, and the relative prevalence of some of them (e.g., cerium) makes them additionally acceptable. The chemical properties of the lanthanides, for example the low solubility of their oxides and hydroxides, make them particularly suitable for corrosion inhibition [28–31]. Excellent corrosion inhibition is attributed to the high reactivity of rare earths. In the articles, the formation of a protective film and the inhibition of the cathodic corrosion reaction have been proven [32–37].

Imidazolines are a special class of corrosion inhibitors that have proven to be highly effective, and therefore, they have a wide range of applications in protection against CO_2

corrosion. In addition to being highly effective in protecting against CO_2 corrosion, they are also available and environmentally friendly. To further improve their effectiveness, commercial forms of imidazolines are most often mixtures of imidazolines and other compounds, i.e., these are imidazoline derivatives [38]. Imidazoline derivatives are organic compounds based on nitrogen that act as cationic surfactants, depending on the nature of the hydrocarbon substituting group attached to the carbon or nitrogen atom of the imidazole ring [39]. The imidazoline head (with free electron pairs on nitrogen atoms and delocalized π-electrons) binds to the metal surface, while the hydrophobic tail represents a kind of barrier against the diffusion of water ions [40]. The inhibitory effect is attributed to physical (electrostatic interaction of the inhibitor molecule and the metal) or chemical adsorption (sharing or transfer of electrons between the inhibitor and the metal) on the metal surface, resulting in the formation of a protective film [39].

The examination of the synergistic effect of the compounds is carried out with the assumption of an increase in the inhibitory effectiveness in corrosion protection, as well as due to a reduction in the doses of chemicals [38,41–46]. The synergistic action involving rare earth metal salts as corrosion inhibitors is rarely investigated; therefore, the aim of this work is to examine the potentially synergistic action of rare earth salt, $CeCl_3$ and imidazoline based compounds.

In this work, the joint/synergistic effect of cerium chloride and imidazoline on the corrosion of carbon steel in a simulated brine solution saturated with CO_2 at 60 °C (which corresponds to the average temperature of the transported fluid in the oil industry) was tested. The inhibitory action of individual components as well as their synergistic action was tested using the mass loss method as well as the electrochemical methods of potentiodynamic polarization, measurement of polarization resistance and electrochemical impedance spectroscopy. The surface analysis of carbon steel samples was examined using scanning electron microscopy with energy-dispersive spectroscopy, X-ray diffraction and Raman spectroscopy.

2. Materials and Methods

The tested alloy is the AISI 1018 carbon steel with the chemical composition determined via optical emission spectrometry on a GDS 850 Leco spectrophotometer, MI, USA, and it is composed of the following (wt.%): 0.35 C, 0.35 Si, 1.09 Mn, 0.007 P, 0.024 S, 0.09 Cr, 0.07 Ni, 0.01 Mo, 0.10 Cu, <0.01 V and Fe balance. The steel has a ferrite–pearlite microstructure, as shown in Figure 1.

The carbon steel surface area for tests using the mass loss method was 23.72 cm^2, while for electrochemical tests, the surface area exposed to the action of the tested electrolyte was 1 cm^2. Before testing, the carbon steel samples were wet sanded with a silicon carbide abrasive paper up to 1200 grit, washed with double distilled water, degreased in 96% ethanol and dried in a stream of air.

The tested electrolyte is composed of 800 mL of chloride–carbonate solution (synthetic brine solution, SBS) saturated with CO_2, both without and with inhibitors (individually and in a mixture). The chloride–carbonate solution was prepared by dissolving 30.0 g dm^{-3} NaCl, 0.1 g dm^{-3} $NaHCO_3$ and 0.1 g dm^{-3} $CaCO_3$ in distilled water. The chloride–carbonate solution was saturated with CO_2 by bubbling CO_2 for 1 h before the start of the measurement and continued during the experiment. The composition of the tested electrolyte corresponds to the composition of natural, stratified waters of oil deposits in Croatia, except that the chloride content has been increased to carry out corrosion tests under severe conditions. The pH value of the prepared solution was 5.3. The pH value was measured with a Portable Waterproof pH meter HI991300, Hanna Instruments, Bedfordshire, UK, (pH accuracy ±0.1 pH, pH range 0.0–14.0). The chemicals used for the preparation of simulated brine solution were analytical grade and produced by Sigma Aldrich, Darmstadt, Germany. In the experiment, imidazoline oleic acid (IOA), produced by Sigma Aldrich, Darmstadt, Germany, and cerium chloride, also produced by Sigma Aldrich, Darmstadt, Germany, as corrosion inhibitors were used. The chemical structures of imidazoline based on oleic acid and cerium chloride are shown in Figure 2.

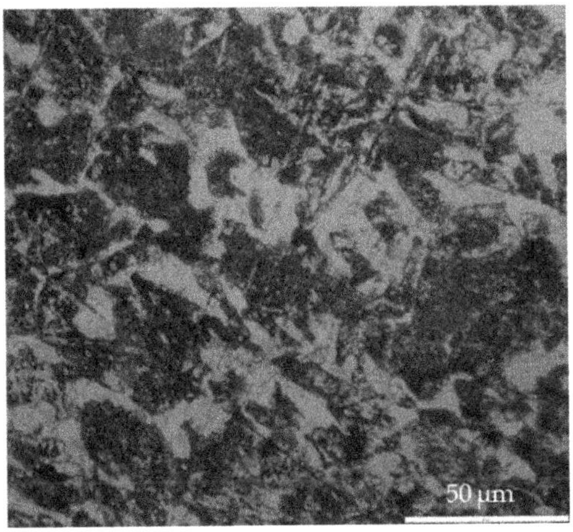

Figure 1. The ferrite–pearlite microstructure of the tested AISI 1018 carbon steel scanned with an Olympus GX51 light microscope, Tokyo, Japan, magnification 1000×.

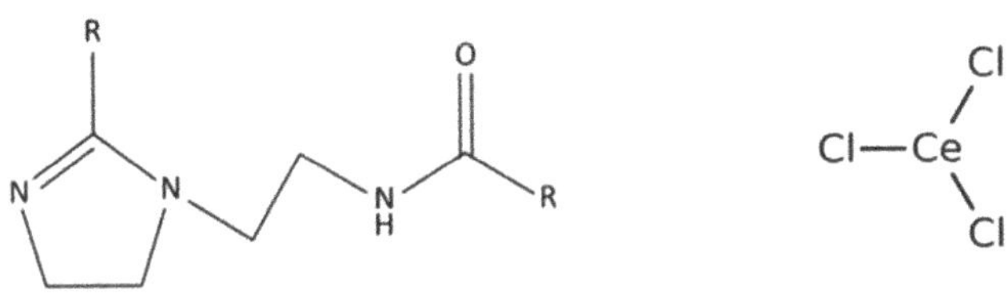

Figure 2. Chemical structures of (**a**) oleic acid imidazoline-based compound and (**b**) cerium chloride anhydrate.

Both tested compounds were added individually in concentrations of 5 mg dm^{-3}, 10 mg dm^{-3}, 15 mg dm^{-3} and 20 mg dm^{-3}, while the inhibitor mixture consisted of 5 mg dm^{-3}, 10 mg dm^{-3}, 15 mg dm^{-3} or 20 mg dm^{-3} IOA with the addition of 5 mg dm^{-3}, 10 mg dm^{-3} or 15 mg dm^{-3} $CeCl_3$ to each of these concentrations. The concentrations of individual components used in this work are arbitrary and lower than those reported in the literature for similar test systems. The aim in selecting concentrations was to achieve satisfactory efficacy at as low a concentration of individual components and as low a concentration of their mixtures as possible, i.e., use the lowest possible doses of chemicals.

All measurements were carried out at a temperature of 60 ± 1 °C with stirring at a speed of 300 rpm; they were repeated at least three times, and very good reproducibility of the results was achieved (relative standard deviation was less than 5%).

2.1. Mass Loss Measurements

Tests using the mass loss method were performed in accordance with the ASTM G-1 standard. Before immersion in the tested electrolyte, the steel coupons were prepared in the previously described manner (treated with sandpaper, washed, degreased and dried) and weighed on an analytical balance with an accuracy of ±0.1 mg. After 72 h of exposure to the tested electrolytes, the coupons were taken out, and their surfaces were cleaned with a Clarke's solution (ASTM G 1-03 pickling solution consists of 5% stannous chloride and 2% antimony trioxide in concentrated hydrochloric acid), washed with distilled water, degreased with 96% ethanol, dried in a stream of air and weighed again on an analytical balance. The chemicals used for the preparation of Clarke's solution and the ethanol used for degreasing were analytical grade and produced by Sigma Aldrich, Darmstadt, Germany. All tests were performed at least four times, and very good reproducibility of the obtained measurement results was achieved.

2.2. Electrochemical Measurements

Electrochemical corrosion tests were performed on an EG&G PAR 378 potentiostat/galvanostat, IL, USA, in conjunction with the Power Suite software 2.41. An electrochemical cell with a three-electrode system was connected to the potentiostat. The working electrode was a sample of tested carbon steel AISI 1018, with a circular cross-section, inserted into the holder of the working electrode, so that the surface of the steel exposed to the action of the electrolyte was 1 cm^2. A saturated calomel electrode was used for the reference electrode, and two graphite rods were used for the counter electrodes.

The electrochemical measurements were performed after 24 h of stabilization of the working electrode at the open circuit potential.

Linear polarization curves in a narrow range of potential were recorded, ±10 mV in relation to the open circuit potential with a scanning rate of 0.1660 mV s^{-1}, from which the polarization resistances were determined.

Potentiodynamic polarization recordings were performed in a potential range of ±250 mV in relation to the open circuit potential with a scanning rate of 0.1660 mV s^{-1}. Corrosion rates were calculated using the Tafel extrapolation method. The method includes extrapolation of the linear cathodic and linear anodic parts of the polarization curve to their intersection, which corresponds to the corrosion current and corrosion potential from which the software calculates the corrosion rates.

Electrochemical impedance spectra were recorded at an open circuit potential, in the frequency range from 100 kHz to ~10 mHz and an excitation signal of 5 mV. The ZSimpWin software 3.21 was used to analyze the recorded impedance spectra.

2.3. Scanning Electron Microscopy (SEM) and Energy-Dispersive X-Ray Spectroscopy (EDS) Analyses

The morphology of the carbon steel surface products after 72 h of exposure to systems with single components and their mixtures was examined by a scanning electron microscope (SEM) JEOL JSM6510LV, Tokyo, Japan, while the chemical properties of the surface products were determined via energy-dispersive X-ray spectroscopy (EDS). Analyses of the carbon steel were performed after exposure to the system in the presence of individual components separately at concentrations of 15 mg dm^{-3} $CeCl_3$ and 15 mg dm^{-3} IOA and in the presence of a mixture of 15 mg dm^{-3} $CeCl_3$ and 15 mg dm^{-3} IOA. This combination of the compounds was chosen because most of the test methods showed the highest inhibitory effectiveness at precisely this concentration of the mixture. The samples analyzed were previously sputtered with gold.

2.4. Raman Spectroscopy

The Raman spectra of the carbon steel surface corrosion products were obtained using a Bruker Senterra II confocal Raman microscope, Karlsruhe, Germany, with a 50× long working distance objective. Raman excitation was performed with a Torus single frequency of 532 nm, continuous-wave laser at a laser power of 25 mW and a laser spot diameter of

2–3 µm. Spectra were collected in the 50–4250 cm^{-1} wavenumber range using a grating with 1200 lines/mm, with an acquisition time of 30 s. The calibration was automatic using SureCAL technology, Brucker, Karlsruhe, Germany. The peak positions were obtained using curve fitting in the Opus software (Base Package).

The samples were analyzed in their original form, without prior preparation, after 72 h of exposure to systems with single components (15 mg dm^{-3} $CeCl_3$; 15 mg dm^{-3} IOA) and their mixtures (of 15 mg dm^{-3} $CeCl_3$ and 15 mg dm^{-3} IOA).

2.5. Powder X-Ray Diffraction Analysis (PXRD)

The samples of pure carbon steel AISI 1018 and those of carbon steel after 72 h of exposure to the synthetic brine solution saturated with CO_2 and with the addition of 15 mg dm^{-3} $CeCl_3$ were analyzed using the powder method on a Panalytical Empyrean diffractometer, Malvern, UK. Copper radiation (Kα 1.541874 Å) was used for the analysis with the following recording conditions: U = 45 kV, I = 40 mA, scan area 3–70° 2Θ, step height 0.0131°, scan rate 0.041683°/s, primary beam divergence 1/2° and PIXcel3D-Medipix3 1 × 1 line detector. Recorded diffractograms were interpreted using the High Score Plus 4.9 (2020) computer program, which uses the PDF-2 database.

3. Results and Discussion

3.1. Mass Loss Measurement Method

The results of tests using the mass loss method are shown in Table 1 (values of corrosion rates, v_{corr}, inhibitor effectiveness, IE, for the uninhibited system and for systems in the presence of individual compounds and their mixtures at different concentrations, γ, and the synergistic parameter, S).

Table 1. Results obtained using the mass loss method.

System	γ (mg L^{-1})	v_{corr} (mm $year^{-1}$)	IE (%)	S
Synthetic brine solution	0	1.57	/	/
IOA	5	1.02	35.03 ± 0.02	/
	10	0.58	63.06 ± 0.51	
	15	0.49	68.79 ± 0.49	
	20	0.17	89.17 ± 0.92	
$CeCl_3$	5	0.53	66.24 ± 0.03	/
	10	0.52	66.88 ± 0.15	
	15	0.53	66.24 ± 0.26	
	20	0.56	64.33 ± 0.18	
$CeCl_3$ + IOA	5 + 5	0.29	81.53 ± 0.90	1.17
	5 + 10	0.22	85.99 ± 0.94	0.87
	5 + 15	0.26	83.43 ± 0.88	0.62
	5 + 20	0.14	91.08 ± 1.02	0.40
$CeCl_3$ + IOA	10 + 5	0.13	91.72 ± 1.12	2.60
	10 + 10	0.14	91.08 ± 0.98	1.37
	10 + 15	0.13	91.72 ± 1.04	1.27
	10 + 20	0.12	92.36 ± 1.22	0.47
$CeCl_3$ + IOA	15 + 5	0.11	92.99 ± 1.15	3.13
	15 + 10	0.08	94.90 ± 1.45	2.45
	15 + 15	0.05	96.82 ± 1.86	2.76
	15 + 20	0.06	96.18 ± 1.88	1.15

The corrosion rate, v_{corr}, was calculated according to Equation (1):

$$v_{corr}\left(\text{mm year}^{-1}\right) = \frac{(m_1 - m_2)}{t\rho A} \quad (1)$$

where m_1 and m_2 are the average mass of samples before and after exposure, respectively; A is the surface of the carbon steel exposed to the electrolyte; t is the exposure time, and ρ is the iron density.

The effectiveness of the inhibitor was calculated from the corrosion rates according to Equation (2):

$$IE\ (\%) = \left(\frac{v^0_{corr} - v^{inh}_{corr}}{v^0_{corr}}\right) \times 100 \quad (2)$$

where v^0_{corr} is the corrosion rate of carbon steel in the system without inhibitor, and v^{inh}_{corr} is the corrosion rate of carbon steel in the system with individual inhibitor or in the presence of individual inhibitors' mixtures.

The measured mass loss, i.e., the calculated corrosion rate of carbon steel in a chloride–carbonate solution saturated with CO_2 at 60 °C is $v^{inh}_{corr} = 1.57$ mm year^{-1}, is in accordance with literature data [47].

The corrosion rate of carbon steel in the presence of cerium(III) chloride is lower (from 0.52 mm year^{-1} to 0.56 mm year^{-1}) compared to the uninhibited system (1.57 mm year^{-1}); however, the corrosion rates and inhibitor effectiveness do not change significantly with increasing $CeCl_3$ concentration (v^{inh}_{corr} from 0.52 mm year^{-1} to 0.56 mm year^{-1}, i.e., IE from 64.24% to 66.88%). The lowest corrosion rate of carbon steel, i.e., the highest effectiveness with the addition of $CeCl_3$, was achieved at a concentration of 10 mg/L (v_{corr} = 0.52 mm year^{-1}, IE = 66.88 ± 0.15%).

The system in the presence of an imidazoline-based oleic acid inhibitor showed a decrease in the corrosion rate (from 1.02 mm year^{-1} to 0.17 mm year^{-1}) and an increase in effectiveness (from 35.03% to 89.17%) with an increase in the concentration of the inhibitor (from 5 mg dm^{-3} to 20 mg dm^{-3}). The highest inhibitory efficiency, IE = 89.17%, was achieved via IOA at a concentration of 20 mg/L.

The synergistic effect of the mixture of two compounds, IOA and $CeCl_3$, as corrosion inhibitors is expressed through a synergistic parameter, S, which is determined according to Equation (3) [48,49]:

$$S = \frac{1 - IE_{IOA} - IE_{CeCl_3} + IE_{IOA} \cdot IE_{CeCl_3}}{1 - IE_{IOA+CeCl_3}} \quad (3)$$

where IE_{IOA} and IE_{CeCl_3} are the inhibition efficiencies of individual compounds, respectively, while $IE_{IOA+CeCl_3}$ is the inhibition efficiency of a mixture of IOA and $CeCl_3$. Values of the synergistic parameter greater than one indicate a synergistic effect of the compounds, while values less than one indicate their antagonistic action [49,50].

Mixtures of compounds IOA and $CeCl_3$, in most of combinations, show synergistic action ($S > 1$) and higher inhibitory effectiveness (from 81.53% to 96.82%) compared to individual components (IE is from 64.24% to 66.88% for $CeCl_3$ and from 35.03% to 89.17% for IOA). The highest inhibitory efficiency, IE = 96.82%, was achieved at a mixture concentration of 15 mg/dm^3 $CeCl_3$ and 15 mg/dm^3 IOA.

3.2. Results of Electrochemical Measurements

3.2.1. Linear Polarization in a Narrow Range of Potential, $E_{corr} \pm 10$ mV

Figure 3 shows the values of polarization resistances, R_p, with relative standard deviation (less than 5%) determined from the recorded linear polarization curves in a narrow potential range, $E_{corr} \pm 10$ mV, for different concentrations of individual components ($CeCl_3$ and IOA) and their mixtures after 24 h of exposure to tested electrolyte.

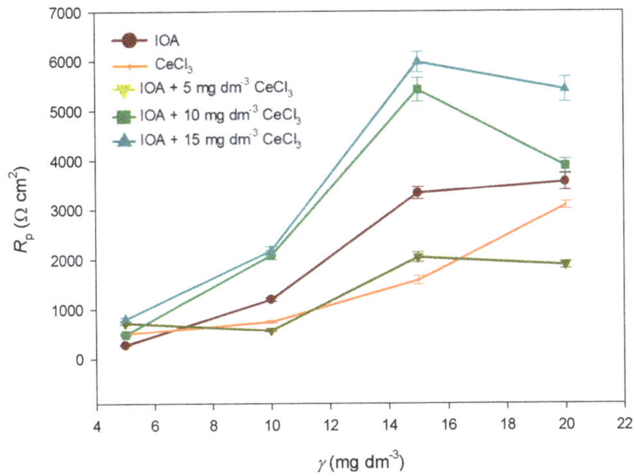

Figure 3. Polarization resistance values, R_p, with relative standard deviation, determined from recorded linear polarization curves in a narrow potential range, $E_{corr} \pm 10$ mV, for various concentrations of individual components ($CeCl_3$ and IOA) and their mixtures.

The values of electrochemical kinetic parameters (j_{corr}—corrosion current density, v_{corr}—corrosion rate, R_p—polarization resistance, *IE*—inhibitory efficiency and *S*—synergistic parameter) determined from linear polarization curves and recorded on the AISI 1018 carbon steel in a narrow potential range, $E_{corr} \pm 10$ mV, after 24 h exposure to the tested electrolyte, are shown in Table 2.

In all tested systems, in the presence of individual components, $CeCl_3$ and IOA as well as their mixtures, a shift in corrosion potentials in the positive direction, an increase in polarization resistance values and a decrease in corrosion rates compared to the blank system were observed. The efficacies of the tested compounds as corrosion inhibitors, individually and in their mixture, were calculated from the corrosion rate values according to Equation (2). The synergistic effect of each compound's mixture was calculated according to Equation (3).

In the case of IOA, an increase in polarization resistance, from 275.1 Ω cm^{-2} to 3555.1 Ω cm^{-2}, with an increase in concentrations from 5 mg dm^{-3} to 20 mg dm^{-3} of IOA is observed. The corrosion rate decreases from 1.115 mm $year^{-1}$ to 0.077 mm $year^{-1}$, and the effectiveness of IOA as a corrosion inhibitor increases from 11.8% to 92.89%. IOA achieved its highest effectiveness, 92.89%, at a concentration of 20 mg dm^{-3}.

The values of polarization resistances, R_p, for $CeCl_3$ show an increase (from 508.4 Ω cm^{-2} to 3082.9 Ω cm^{-2}) with an increasing concentration (from 5 mg dm^{-3} to 20 mg dm^{-3}). The corrosion rates range from 0.232 mm $year^{-1}$ to 0.055 mm $year^{-1}$, and the effectiveness of $CeCl_3$ as a corrosion inhibitor increases from 81.66% to 95.65%. In the system with $CeCl_3$ as a corrosion inhibitor, the highest efficiency, 95.65%, was achieved at the highest concentration of the compound, 20 mg dm^{-3}. Compared to the action of IOA alone, $CeCl_3$ alone shows greater effectiveness in protecting AISI 1018 carbon steel from corrosion under the tested conditions.

Table 2. Electrochemical kinetic parameters determined from linear polarization curves and recorded on the AISI 1018 carbon steel after 24 h exposure to the tested electrolyte.

System	γ (mg dm^{-3})	E_{corr} (mV)	j_{corr} (µA cm^{-2})	R_p (Ω cm^{-2})	v_{corr} (mm year^{-1})	*IE* (%)	S
SBS	0	−718	109.0	252.9	1.265	/	/
IOA	5	−709	98.86	275.1	1.115	11.86 ± 0.04	/
	10	−672	24.28	1195.2	0.282	77.71 ± 0.38	
	15	−645	8.23	3329.8	0.096	92.41 ± 1.17	
	20	−644	6.34	3555.1	0.077	92.89 ± 0.95	
$CeCl_3$	5	−685	18.00	508.4	0.232	81.66 ± 0.79	/
	10	−679	15.78	745.9	0.183	85.53 ± 0.53	
	15	−690	10.10	1577.2	0.117	90.75 ± 1.32	
	20	−696	4.76	3082.9	0.055	95.65 ± 0.71	
$CeCl_3$ + IOA	5 + 5	−662	12.40	727.1	0.144	88.62 ± 0.57	1.42
	5 + 10	−675	17.78	565.9	0.206	83.72 ± 1.31	0.16
	5 + 15	−657	7.33	2040.9	0.085	93.28 ± 1.05	0.13
	5 + 20	−656	7.10	1883.1	0.082	93.52 ± 1.30	0.04
$CeCl_3$ + IOA	10 + 5	−669	22.76	490.1	0.264	79.13 ± 1.44	0.61
	10 + 10	−639	7.37	2085.1	0.085	93.28 ± 1.99	0.48
	10 + 15	−636	3.17	5417.4	0.037	97.08 ± 1.43	0.24
	10 + 20	−642	3.89	3880.9	0.045	96.44 ± 0.93	0.03
$CeCl_3$ + IOA	15 + 5	−658	9.29	807.4	0.108	91.46 ± 1.35	0.95
	15 + 10	−648	4.73	2171.5	0.055	95.65 ± 2.03	0.47
	15 + 15	−634	2.24	5975.3	0.026	97.94 ± 0.45	0.53
	15 + 20	−635	3.40	5415.5	0.039	96.92 ± 0.28	0.06

Mixtures of 5 mg dm^{-3} $CeCl_3$ and IOA (from 5 mg dm^{-3} to 20 mg dm^{-3}) do not show an increase in polarization resistance compared to the individual components. However, at higher concentrations of $CeCl_3$ (10 mg dm^{-3} and 15 mg dm^{-3}) in mixtures with IOA (5 mg dm^{-3}, 10 mg dm^{-3}, 15 mg dm^{-3} and 20 mg dm^{-3}), an increase in polarization resistances is observed compared to the individual components. In mixtures of 15 mg dm^{-3} $CeCl_3$ added to IOA, the highest values of polarization resistance were achieved. Tests at the highest concentrations of the tested mixtures (20 mg dm^{-3} IOA in a mixture with 5 mg dm^{-3}, 10 mg dm^{-3} and 15 mg dm^{-3} $CeCl_3$) show a decrease in resistance compared to the values of polarization resistances achieved at previous lower concentrations. A possible explanation is that for organic compounds, which IOA is, a critical micellar concentration has been reached [51]. The lowest corrosion rate, 0.026 mm year^{-1}, and the highest inhibitory efficacy, 97.94%, were achieved in the presence of a mixture of 15 mg dm^{-3} $CeCl_3$ and 15 mg dm^{-3} IOA. Although most combinations of $CeCl_3$ and IOA mixtures show better performance in corrosion protection of AISI 1018 carbon steel under the tested conditions, compared to the individual components, a synergistic effect ($S > 1$) was demonstrated only for the 5 mg dm^{-3} $CeCl_3$ and 5 mg dm^{-3} IOA combination.

3.2.2. Potentiodynamic Polarization with Tafel Extrapolation

Polarization curves recorded on carbon steel AISI 1018 in a chloride–carbonate solution saturated with CO_2 at 60 °C without the addition of inhibitors, in the presence of individual components and their mixtures, are shown in Figures 4 and 5.

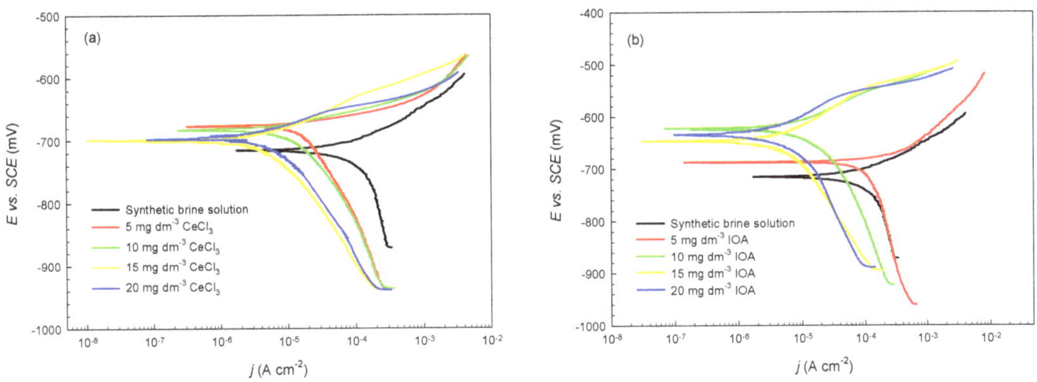

Figure 4. Polarization curves recorded on carbon steel AISI 1018 in a chloride–carbonate solution saturated with CO_2 at 60 °C without the addition of inhibitors and in the presence of different concentrations of individual components. (**a**) $CeCl_3$ and (**b**) IOA after 24 h of exposure to the tested electrolyte.

Electrochemical kinetic parameters (j_{corr}—corrosion current density, v_{corr}—corrosion rate, $-\beta_c$ and β_a—cathodic and anodic Tafel slope, respectively, *IE*—inhibitory efficiency and *S*—synergistic parameter) determined from the polarization curves are summarized in Table 3. The inhibitory efficacies of individual compounds and their mixtures were calculated according to Equation (2), and the synergistic effects of mixtures of compounds were calculated according to Equation (3).

Table 3. Electrochemical kinetic parameters determined from the potentiodynamic polarization curves recorded in a potential range of E_{corr} ± 250 mV relative to the open circuit potential.

System	γ (mg dm^{-3})	E_{corr} (mV)	j_{corr} (μA cm^{-2})	$-\beta_c$ (mV dec^{-1})	β_a (mV dec^{-1})	v_{corr} (mm year^{-1})	*IE* (%)	S
SBS	0	−713	124.6	410.69	75.01	1.446	/	/
IOA	5	−689	117.0	406.04	68.30	1.358	6.09 ± 0.01	/
	10	−623	15.75	229.49	94.09	0.183	87.34 ± 0.25	
	15	−625	6.72	200.76	91.94	0.078	94.61 ± 0.95	
	20	−582	11.42	152.81	91.62	0.133	90.80 ± 0.87	
$CeCl_3$	5	−677	17.87	211.92	23.37	0.207	85.68 ± 0.56	/
	10	−684	14.55	171.32	32.15	0.169	88.31 ± 0.68	
	15	−700	4.67	144.11	49.16	0.054	96.27 ± 0.94	
	20	−694	4.76	117.50	47.30	0.055	96.19 ± 1.03	
$CeCl_3$ + IOA	5 + 5	−641	15.62	133.12	24.57	0.181	87.48 ± 0.75	1.07
	5 + 10	−658	13.78	112.01	29.18	0.159	89.00 ± 0.67	0.14
	5 + 15	−650	3.54	83.68	58.37	0.041	97.17 ± 1.24	0.07
	5 + 20	−654	3.54	78.13	50.71	0.041	97.17 ± 1.35	0.12
$CeCl_3$ + IOA	10 + 5	−660	29.91	245.85	28.64	0.347	76.00 ± 0.35	0.46
	10 + 10	−625	5.08	76.05	65.99	0.059	95.91 ± 0.77	0.36
	10 + 15	−626	3.70	96.94	66.75	0.043	97.03 ± 0.69	0.36
	10 + 20	−625	3.70	92.44	78.31	0.043	97.03 ± 0.73	0.36
$CeCl_3$ + IOA	15 + 5	−639	6.87	77.62	22.18	0.080	94.47 ± 1.02	0.63
	15 + 10	−630	5.83	101.39	30.79	0.067	95.37 ± 1.11	0.10
	15 + 15	−625	2.74	63.49	59.50	0.032	97.78 ± 0.78	0.09
	15 + 20	−629	3.21	84.60	58.80	0.037	97.44 ± 0.25	0.13

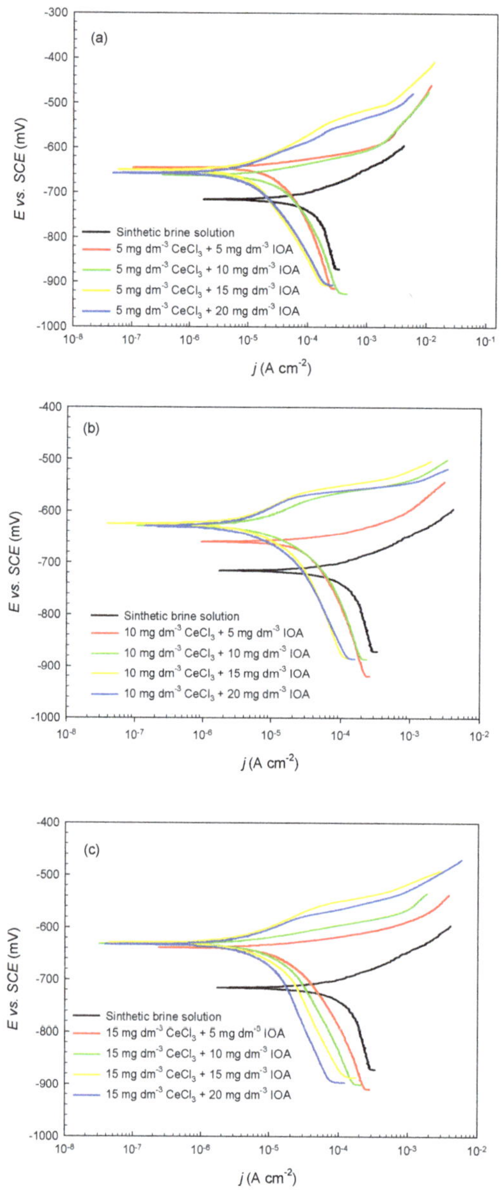

Figure 5. Polarization curves recorded on carbon steel AISI 1018 in a chloride–carbonate solution saturated with CO_2 after 24 h of exposure to the tested electrolyte at 60 °C without the addition of inhibitors and in the presence of different compounds' mixtures: (**a**) 5 mg dm^{-3} $CeCl_3$ and different concentrations of IOA, (**b**) 10 mg dm^{-3} $CeCl_3$ and different concentrations of IOA and (**c**) 15 mg dm^{-3} $CeCl_3$ and different concentrations of IOA.

The mechanism of CO_2 carbon steel corrosion has been a controversial subject for decades. The system is very complex, involving many chemically and electrochemically active species in an acidic CO_2/H_2O system. CO_2 dissolves in water, which then reacts chemically with water to form a weak carbonic acid, H_2CO_3, which dissociates in two steps. In the first step, it dissociates into a hydrogen ion, H^+, and a bicarbonate ion, HCO_3^-, which in the second step, dissociates into a hydrogen ion, H^+, and a carbonate ion, CO_3^{2-}, according to the following reactions:

$$CO_{2\ (g)} \leftrightarrow CO_{2\ (aq)} \tag{4}$$

$$CO_{2\ (ag)} + H_2O_{(l)} \leftrightarrow H_2CO_{3\ (aq)} \tag{5}$$

$$H_2CO_{3\ (ag)} \leftrightarrow HCO_3^-{}_{(aq)} + H^+{}_{(aq)} \tag{6}$$

$$HCO_3^-{}_{(aq)} \leftrightarrow CO_3^{2-}{}_{(aq)} + H^+{}_{(aq)} \tag{7}$$

Water as a solvent can dissociate according to the following reaction:

$$H_2O_{\ (aq)} \leftrightarrow OH^-{}_{(aq)} + H^-{}_{(aq)} \tag{8}$$

Under the tested conditions, the anodic corrosion reaction is the oxidation of iron from steel according to the following reaction:

$$Fe_{(s)} \leftrightarrow Fe^{2+}{}_{(aq)} + 2e^- \tag{9}$$

Cathodic reactions of CO_2 corrosion involve the reduction of the hydrogen ions, H^+, either directly or via other weak acids present in the system (H_2CO_3, HCO_3^-, H_2O), according to the following reactions [52]:

$$2H^+{}_{(aq)} + 2e^- \leftrightarrow H_{2\ (g)} \tag{10}$$

$$2H_2O_{\ (l)} + 2e^- \leftrightarrow 2OH^-{}_{(aq)} + H_{2\ (g)} \tag{11}$$

$$2H_2CO_{3\ (aq)} + 2e^- \leftrightarrow 2HCO_3^-{}_{(aq)} + H_{2\ (g)} \tag{12}$$

$$2HCO_3^-{}_{(aq)} + 2e^- \leftrightarrow 2CO_3^{2-}{}_{(aq)} + H_{2\ (g)} \tag{13}$$

It was found that under the test conditions analogous to those in this work, carbonic acid affects the limiting cathodic current but has no effect on the charge transfer current. The charge transfer current depends only on pH, indicating that hydrogen ion reduction is the main cathodic reaction (this mechanism is called the "buffering effect") according to Reaction 10. The direct reduction of carbonic acid is insignificant compared to the reduction of hydrogen ions [52,53].

The polarization curve recorded on carbon steel after 24 h of exposure to the inhibitor-free system (black curve in Figures 4 and 5) shows cathodic corrosion currents partly influenced by mass transfer, while the anodic polarization curves show a linear dependence of current and potential, indicating a region of active dissolution, i.e., iron oxidation [54].

In the presence of IOA, an inhibitory effect on the cathodic and anodic process was observed, i.e., a shift in the cathodic and anodic polarization curves towards lower values of the corrosion current density and thus lower corrosion rates. Such an inhibitory action is more pronounced with an increase in the concentration of IOA, which is evident from the polarization curves (Figure 4) as well as the kinetic parameters contained in Table 3. The slopes of the anodic Tafel lines in the presence of IOA change slightly (68.30 mV dec^{-1}–94.09 mV dec^{-1}) compared to the value of the slope in the uninhibited solution (75.01 mV dec^{-1}), which points to the effect of IOA in the way that it adsorbs on the metal surface, blocks the active sites and prevents the anodic reaction from taking place rather than changing the mechanism of the anodic reaction [13,14,51]. The dominant cathodic reduction reaction in the test conditions (CO_2-saturated brine solution, pH $\approx$ 5) is the reduction of hydrogen ions produced via the dissociation of carbonic acid. The slope of the cathodic Tafel line of the uninhibited system is high (410.69 mV dec^{-1}) in electrolytes due to the influence of diffusion of H^+ ions on the cathodic reaction. In the presence of IOA, a decrease in the slope of the cathodic Tafel line (406.04 mV dec^{-1}–152.81 mV dec^{-1}) is observed, but it is still high, indicating the influence of diffusion on the cathodic process [15]. In the presence of IOA, the corrosion potential is shifted towards more positive values (from −689 mV to −582 mV), which indicates a preference for inhibiting the anodic reaction and is associated with the formation of a protective film [17,51]. At higher concentrations of IOA (from 10 mg dm^{-3} to 20 mg dm^{-3}), the corrosion potential shift in the anodic direction is greater than 85 mV compared to the system without inhibitors, classifying it as an anodic type of corrosion inhibitor under the tested conditions [18–23,51]. It slows down corrosion by reducing the rate of the anodic metal dissolution process. The values of inhibitory effectiveness, shown in Table 2, increase with the increase in IOA concentration and reach a maximum value of 94.61% at a concentration of 15 mg dm^{-3}, which is attributed to the binding of imidazoline molecules to the metal surface. The structure of IOA (Figure 2a) shows the existence of delocalized electrons in the imidazole ring as well as a free electron pair on the nitrogen atoms. The adsorption of IOA on the surface of carbon steel AISI 1018 can be explained in the way that IOA can donate these π-electrons and free electron pairs to the empty d-orbitals of iron [55,56].

The presence of $CeCl_3$ causes a shift in the corrosion potential slightly in the positive direction (from −677 mV to −699 mV) compared to the blank system (−713 mV). At all tested $CeCl_3$ concentrations, a displacement of the cathodic and anodic branches of the polarization curves towards lower corrosion currents is observed. The corrosion current densities (from 17.87 μA cm^{-2} to 4.67 μA cm^{-2}) and corrosion rates (from 0.207 mm $year^{-1}$ to 0.054 mm $year^{-1}$) are significantly lower compared to the values recorded for the same concentrations in the presence of IOA (from 117.0 μA cm^{-2} to 6.72 μA cm^{-2} and from 1.358 mm $year^{-1}$ to 0.078 mm $year^{-1}$) and the blank system (124.6 μA cm^{-2}, 1.446 mm $year^{-1}$). $CeCl_3$ significantly reduced the slope of the cathodic Tafel line (from 211.92 mV dec^{-1} to 117.50 mV dec^{-1}) compared to the blank system (410.69 mV $year^{-1}$), as shown in Figure 4 and Table 3, indicating the effect of $CeCl_3$ on the cathodic reaction. The sustained high slopes of the cathodic Tafel lines indicate the influence of diffusion on the cathodic reaction [15]. The slopes of the anodic Tafel line are significantly lower (from 47.30 mV dec^{-1} to 23.37 mV dec^{-1}) compared to the uninhibited system (75.01 mV dec^{-1}), indicating an influence of $CeCl_3$ on the anodic corrosion reaction of ferrous oxidation. The effectiveness of $CeCl_3$ in protecting AISI 1018 carbon steel from corrosion increases with increasing concentration (from 5 mg dm^{-3} to 15 mg dm^{-3}). However, a deviation from this trend was observed at the highest tested concentration (20 mg dm^{-3}), where a slight increase in the corrosion rate and a decrease in inhibitory efficiency, compared to values at 15 mg dm^{-3} $CeCl_3$, were observed, which can be explained by the increase in chloride concentration when adding a higher concentration of $CeCl_3$ [33,57–59]. $CeCl_3$ showed the behavior of a mixed corrosion inhibitor with a predominant influence on the anodic process. The highest inhibitory efficiency, 96.27%, was shown by $CeCl_3$ at a concentration of 15 mg dm^{-3}, which potentially can be attributed to the formation of cerium carbonate under the tested conditions [57].

The corrosion potential values for all tested compound mixtures show a shift in the anodic direction compared to the blank system (−713 mV). The range of corrosion potential for the compound mixtures ranges from −660 mV to −625 mV; in some mixtures, only a negligible increase in potential (slightly above 85 mV) compared to the uninhibited system was observed. The values of corrosion currents and corrosion rates are very low in the presence of the compounds′ mixtures, which are already at low concentrations of compounds. The slopes of the cathodic and anodic Tafel lines in the presence of the mixtures of compounds are comparable or lower to those determined in the systems with the individual components, indicating the effect of the mixtures of compounds on the cathodic and anodic corrosion reactions. The mixture of tested compounds blocks the metal surface and reduces the areas available for cathodic and anodic reactions to occur [57]. From the above results, the mixture of $CeCl_3$ and IOA compounds showed the activity of a mixed corrosion inhibitor with a significant effect on the anodic reaction of iron oxidation. The highest inhibitory efficiency, 97.78%, was shown by the mixture composed of 15 mg dm^{-3} $CeCl_3$ and 15 mg dm^{-3} IOA. Most of the mixtures of compounds did not show synergistic action, since as individual components, they achieved high inhibitory efficiency. Nonetheless, an enhanced corrosion protection effect was certainly noted compared to individual compounds. Synergistic action (S = 1.07) was proven only for the combination of the lowest concentrations of compounds: 5 mg dm^{-3} $CeCl_3$ and 5 mg dm^{-3} IOA, with a corresponding effectiveness of 87.48%.

At higher concentrations of individual components and their mixtures, an increase in the slope of the anodic polarization curves, immediately after E_{corr}, is observed (seen in Figures 4 and 5), and it is attributed to the adsorption of the inhibitor onto the steel surface. This is followed by an abrupt increase in the corrosion current density with increasing potential, indicating the desorption of the inhibitor. The surface coverage of the adsorbed inhibitor usually decreases with increasing potential because the adsorbed compounds become unstable at high polarization potentials [60].

3.2.3. Electrochemical Impedance Spectroscopy (EIS)

The phenomena at the electrolyte/carbon steel surface interface were investigated using electrochemical impedance spectroscopy. Nyquist and Bode impedance spectra recorded on carbon steel AISI 1018 in a chloride–carbonate solution saturated with and without CO_2 and in the presence of individual components, $CeCl_3$ and IOA, are shown in Figure 6, while spectra recorded in the presence of mixtures of compounds are shown in Figures 7–9.

The recorded impedance spectra for all tested systems are matched to the equivalent electrical circuit models, as shown in Figure 10. EIS parameters determined by that matching are presented in Table 4.

Based on the presented Nyquist and Bode diagrams, the points represent the experimental values, while the lines represent the values of the assumed theoretical model of the electrical equivalent circuit. The results of matching experimental and theoretical values show a very good agreement, with a deviation of less than 5% for each spectrum. Selected theoretical models of electrical circuits, according to the literature, are used to interpret impedance spectra recorded under similar conditions as those listed in this paper [52,61,62].

Three electrical equivalent circuits were used in the analysis of impedance spectra. The circuit shown in Figure 10a was used in the analysis of spectra recorded in the blank system, while the circuits shown in Figure 10b,c were used for spectra recorded in systems containing individual components of $CeCl_3$, IOA and in the presence of their mixtures. The circuit shown in Figure 10c was used in the analysis of the recorded spectra exhibiting an inductive loop, and the circuit shown in Figure 10b was used in the absence of an inductive loop.

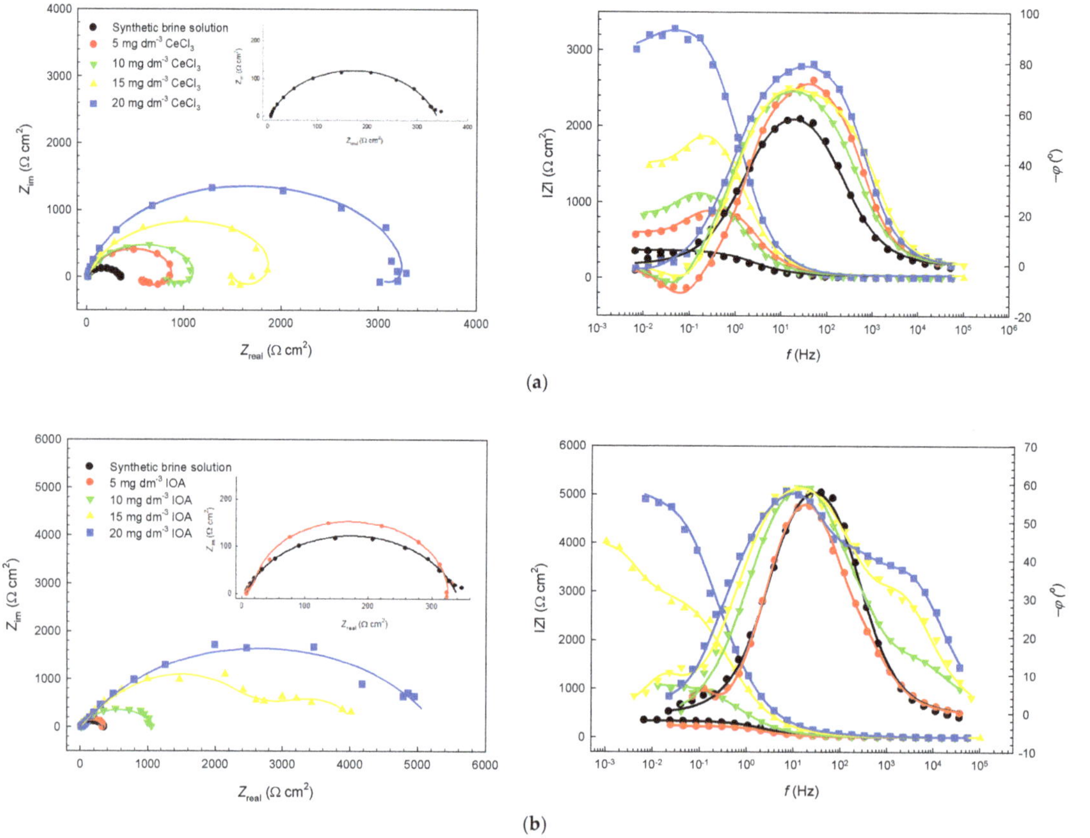

Figure 6. Nyquist and Bode diagrams of impedance spectra recorded on AISI 1018 carbon steel in chloride–carbonate solution saturated with and without CO_2 and in the presence of different concentrations of individual compounds, $CeCl_3$ (**a**) and IOA (**b**) after 24 h of exposure to the tested electrolyte at 60 °C.

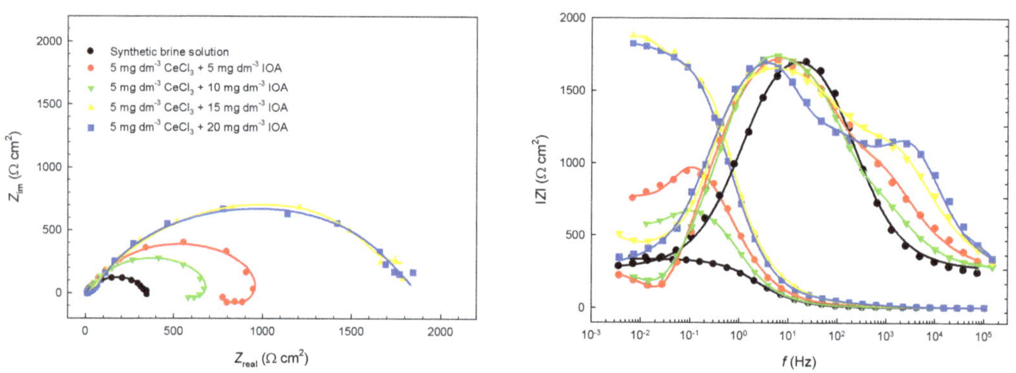

Figure 7. Nyquist and Bode diagrams of impedance spectra recorded on AISI 1018 carbon steel in chloride–carbonate solution saturated with and without CO_2 and in the presence of mixtures of IOA (in concentrations of 5 mg dm^{-3}, 10 mg dm^{-3}, 15 mg dm^{-3} and 20 mg dm^{-3}) and 5 mg dm^{-3} of $CeCl_3$.

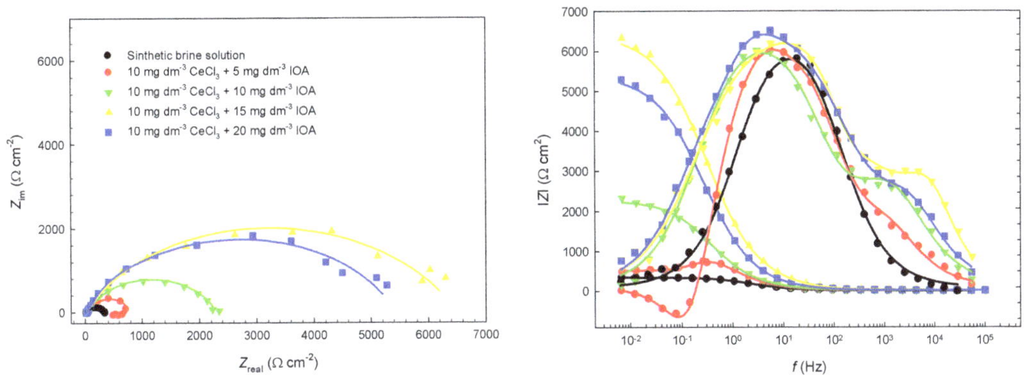

Figure 8. Nyquist and Bode diagrams of impedance spectra recorded on AISI 1018 carbon steel in chloride–carbonate solution saturated with and without CO_2 and in the presence of mixtures of IOA (in concentrations of 5 mg dm^{-3}, 10 mg dm^{-3}, 15 mg dm^{-3} and 20 mg dm^{-3}) and 10 mg dm^{-3} of $CeCl_3$.

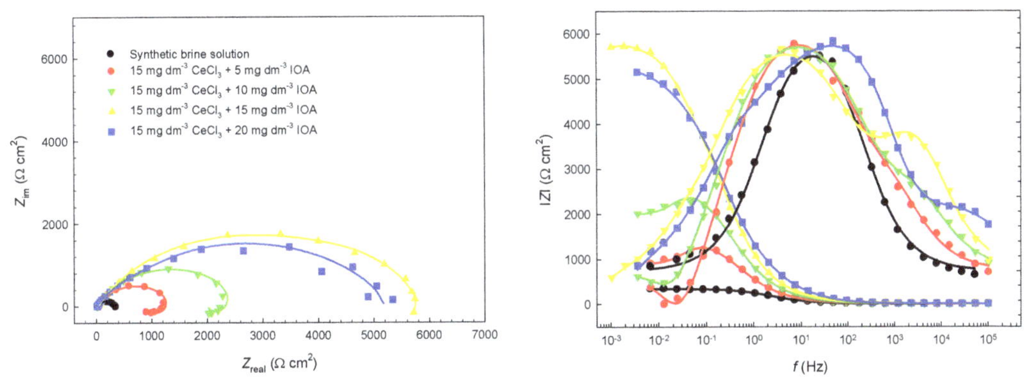

Figure 9. Nyquist and Bode diagrams of impedance spectra recorded on AISI 1018 carbon steel in chloride–carbonate solution saturated with and without CO_2 and in the presence of mixtures of IOA (in concentrations of 5 mg dm^{-3}, 10 mg dm^{-3}, 15 mg dm^{-3} and 20 mg dm^{-3}) and 15 mg dm^{-3} of $CeCl_3$.

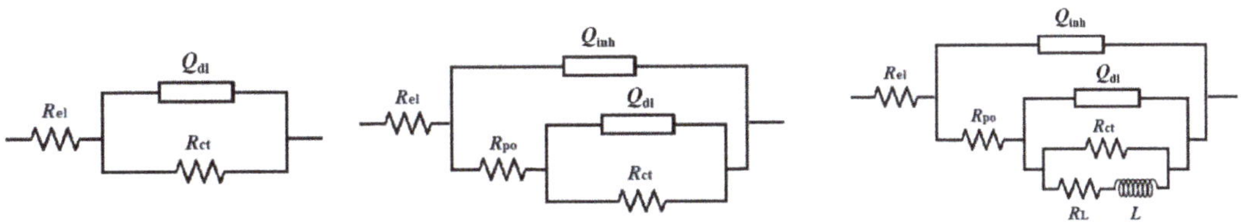

(a) (b) (c)

Figure 10. Equivalent electrical circuits used for the analysis of the recorded impedance spectra of AISI 1018 carbon steel in (**a**) a chloride–carbonate solution without inhibitors and (**b**,**c**) in the presence of individual components ($CeCl_3$ and IOA) and their mixtures.

Table 4. EIS parameters (with chi-square distribution, χ^2) determined by fitting the recorded impedance spectra in the tested electrolytes with the assumed electrical equivalent circuit model.

System	γ (mg L^{-1})	R_{el} (Ω cm^{-2})	Q_{inh} ($\mu S\ s^n$)	R_{po} (Ω cm^{-2})	n_{inh}	Q_{dl} ($\mu S\ s^n$)	R_{ct} (Ω cm^{-2})	n	L (Henri)	R_L (Ω cm^{-2})	IE (%)	S	χ^2 ($\cdot 10^{-4}$)
SBS	0	6.742	/	/	/	1463.7	332.5	0.806	/	/	/	/	1.02
IOA	5	6.889	395.8	15.54	0.826	244.8	344.3	0.826	/	/	3.43 ± 0.01	/	1.01
	10	6.632	104.9	8.467	0.811	286.4	1086.0	0.779	/	/	69.38 ± 0.64		2.74
	15	6.364	287.6	119.7	0.749	167.9	3257.0	0.718	/	/	89.79 ± 0.84		4.86
	20	6.101	95.39	92.81	0.822	86.48	5203.0	0.765	/	/	93.61 ± 0.56		2.61
$CeCl_3$	5	6.593	62.64	102.5	0.892	143.5	783.89	0.968	558.5	330.9	57.58 ± 0.27	/	1,49
	10	6.832	132.4	64.07	0.857	59.81	1159.7	0.794	862.8	327.8	71.33 ± 0.86		3.51
	15	6.867	24.77	27.89	0.848	68.70	2035.0	0.908	803.9	314.3	83.66 ± 0.55		2.44
	20	6.593	56.05	19.90	0.858	45.06	3677.6	0.947	883.1	310.5	90.96 ± 0.78		2.43
$CeCl_3$ + IOA	5 + 5	6.657	179.7	23.91	0.763	189.2	935.2	0.738	944.9	364.5	64.45 ± 0.06	1.15	2.24
	5 + 10	6.202	169.9	22.63	0.737	172.3	1050.1	0.831	877.5	355.2	68.34 ± 0.35	0.41	4.23
	5 + 15	6.688	38.06	21.98	0.747	168.1	1836.9	0.837	/	/	81.90 ± 0.98	0.24	1.77
	5 + 20	6.776	13.86	22.38	0.763	225.9	1829.3	0.905	/	/	81.82 ± 1.12	0.15	3.69
$CeCl_3$ + IOA	10 + 5	6.785	91.96	9.78	0.798	191.3	893.2	0.826	952.6	372.1	62.77 ± 0.69	0.74	2.70
	10 + 10	6.620	80.12	24.52	0.748	237.0	2245.1	0.782	/	/	85.19 ± 0.78	0.59	1.48
	10 + 15	6.618	62.67	12.99	0.735	124.6	6539.5	0.788	/	/	94.92 ± 1.29	0.58	2.16
	10 + 20	6.654	37.02	17.35	0.709	153.0	5836.1	0.739	/	/	94.30 ± 1.32	0.32	3.12
$CeCl_3$ + IOA	15 + 5	6.798	169.0	18.84	0.782	178.4	1424.0	0.776	970.0	354.5	76.65 ± 1.45	0.67	1.53
	15 + 10	6.986	75.95	15.40	0.765	180.9	2711.5	0.747	860.0	228.7	87.74 ± 1.39	0.41	1.17
	15 + 15	6.811	33.61	40.06	0.802	180.3	5831.4	0.722	857.7	209.6	94.30 ± 1.68	0.29	5.07
	15 + 20	6.359	147.8	12.22	0.703	29.32	5435.0	0.785	/	/	93.88 ± 0.29	0.17	1.55

The elements of the chosen electrical equivalent circuits are as follows: electrolyte resistance, R_{el}, charge transfer resistance, R_{ct}, pore resistance of the inhibitor film, R_{po}, inductance, L, resistance of the inductance, R_L, constant phase element equivalent to the capacity of the electrochemical double layer, Q_{dl}, and constant phase element equivalent to the capacity of the inhibitor film, Q_{inh}. A constant phase element, in the electric equivalent circuit, is used instead of a capacitor because it compensates for surface inhomogeneity, geometric irregularity, roughness, and porosity of the electrode, etc., [62,63].

The impedance of a constant phase element is calculated according to

$$Z = \frac{1}{Q}(j\omega)^{-n} \tag{14}$$

where Q is the proportionality factor; $j = \sqrt{-1}$, $\omega = 2\pi f$ and n are the dispersion coefficients associated with surface inhomogeneity. Depending on the value of the coefficient n, the constant phase element can be represented by a resistor (for $n = 0$ $R = Q^{-1}$) or a capacitor (for $n = 1$ $C = Q$). The physical meaning of coefficient n has not been fully clarified [64,65].

The inhibitory efficacy, IE, of individual compounds, $CeCl_3$ and IOA, as well as their mixtures was calculated according to the following equation:

$$IE = \frac{R_{ct}^{inh} - R_{ct}^{0}}{R_{ct}^{inh}} \times 100 \tag{15}$$

where R_{ct}^{inh} represents the charge transfer resistance of carbon steel for the system in the presence of individual $CeCl_3$ and IOA compounds and their mixtures, while R_{ct}^{0} represents the charge transfer resistance for the uninhibited system.

The Nyquist diagram recorded on AISI 1018 carbon steel in a synthetic brine solution saturated with CO_2 and without an inhibitor shows the existence of one capacitive semicircle, as shown in Figure 6. The Bode diagram shows the existence of a single time constant, which indicates an active corrosion process and the formation of corrosion products that do not possess good protective properties [66]. In the analysis of the recorded impedance spectrum for the uninhibited system, the equivalent circuit shown in Figure 10a was used. For similar test conditions, according to the literature, the same equivalent circuit in the analysis of the recorded impedance spectra is used, and by matching the recorded spectrum with the assumed model, a very good agreement was achieved ($\chi^2 = 1.02 \cdot 10^{-4}$). Although the recorded polarization curves for the uninhibited system (Figure 4) show the influence of diffusion on the cathodic reaction, it is not observed in the recorded impedance spectrum. Therefore, the application of the model that includes the Warburg element would result in a large error [67]. The high value of the proportionality factor Q_{dl} = 1463.7 μS s^n is in accordance with the literature data and is consistent with the high value of the double layer capacity in synthetic brine solutions saturated with CO_2 as a result of the large surface area available for the deposition of corrosion products. This can be explained by the ferrite–pearlite microstructure of AISI 1018 carbon steel, where ferrite preferentially dissolves during active corrosion processes while cementite remains, protruding from the steel surface [62,63,68].

In the presence of $CeCl_3$, compared to the blank system, an increase in the real and imaginary impedance components on the Nyquist diagram (Figure 6) is observed. This increase is in line with the increase in $CeCl_3$ concentration, which can be attributed to the formation of a protective film [69]. The recorded spectra show the existence of a small inductive loop in the very-low-frequency range. The inductive loop indicates that the adsorbed inhibitor layer becomes unstable and desorbs from the metal surface, which enhances the relaxation adsorption–desorption processes of the inhibitor. As a result of the destabilization of the adsorbed inhibitory layer, the anodic dissolution of the metal increases [70,71]. In the Bode plots, there is one phase angle peak where the maximum broadens and increases with increasing $CeCl_3$ concentration, which can be attributed to the formation of an inhibitor film on the carbon steel surface [66]. The impedance modulus also increases with increasing $CeCl_3$ concentration. Since the recorded impedance spectra in the presence of $CeCl_3$ show the existence of an inductive loop, the equivalent electrical circuit shown in Figure 10c was used in the analysis. The influence of diffusion, which is visible in the recorded polarization curves in the presence of $CeCl_3$, is not visible in the impedance spectra, and the equivalent circuit with the Warburg element was not used. According to the values in Table 4, the charge transfer resistances, R_{ct}, in the presence of $CeCl_3$ are higher than the charge transfer resistance for the blank system (332.5 Ω cm^2), and they increase (from 783.9 Ω cm^2 to 3677.6 Ω cm^2) with increasing $CeCl_3$ concentration (from 5 mg dm^{-3} to 20 mg dm^{-3}). The values of the constant phase element, Q_{dl}, show lower values (from 143.5 μS s^n to 45.06 μS s^n) compared to the uninhibited system (1463.7 μS s^n), indicating the adsorption of $CeCl_3$ on the steel surface. The inhibitory efficiency, calculated according to Equation (15), increases (from 57.58% to 90.96%) with increasing $CeCl_3$ concentration and achieves the highest value, 90.96%, at 20 mg dm^{-3} $CeCl_3$.

The presence of IOA in the synthetic brine solution on the recorded Nyquist diagrams (Figure 6) is manifested through the increase in capacitive semicircles with increasing IOA concentration. At concentrations above 5 mg dm^{-3} on the Nyquist diagrams, and especially the Bode diagram, in addition to the time constant in the low-frequency range, the appearance of a time constant in the high-frequency region is observed. The maxima of the high-frequency phase angle peaks increase with increasing concentration as do the impedance modules. All the above indicates the formation of an inhibitory film of IOA on the steel surface [39,66]. The electrical equivalent circuit shown in Figure 10b was used in the analysis of the recorded impedance spectra, and a very good agreement was achieved ($\chi^2 = 1.01 \cdot 10^{-4}$–$4.86 \cdot 10^{-4}$). The analysis results, contained in Table 4, show an increase in charge transfer resistance (from 344.3 Ω cm^2 to 5203.0 Ω cm^2) with increasing IOA concentration (from 5 mg dm^{-3} to 20 mg dm^{-3}), which indicates the protective effect of the inhibitor film against carbon steel corrosion under the tested conditions. Lower Q_{dl} values (from 286.4 μS s^n to 86.48 μS s^n) compared to the blank system (1463.7 μS s^n)

indicate the adsorption of IOA onto the carbon steel surface. The decrease in the double layer capacitance, i.e., Q_{dl}, can be explained (via the Helmholtz model) by the replacement of water molecules (high dielectric constant) on the steel surface with inhibitor molecules (low dielectric constant). Another explanation is the increase in the double layer thickness due to the adsorption of larger organic inhibitor molecules on the steel surface, which results in a decrease in the double layer capacitance [72,73]. The effectiveness of IOA as a corrosion inhibitor increases with increasing concentration (from 3.43% to 93.61%) and shows the highest effectiveness, 93.6%, at a concentration of 20 mg dm^{-3}.

The impedance spectra (Nyquist and Bode plots) recorded for mixtures of $CeCl_3$ and IOA compounds are shown in Figures 7–9, while the results of the fitting with equivalent electrical circuits (shown in Figure 10b,c) are contained in Table 4. The recorded impedance spectra in the presence of compound mixtures, in some combinations of concentrations, show the existence of an inductive loop. For them, the best fitting is achieved with the model shown in Figure 10c. For those combinations of compound mixtures, whose spectra do not show the existence of an inductive loop, the best fitting is achieved with the equivalent circuit model shown in Figure 10b. For all tested combinations of $CeCl_3$ and IOA mixtures, a single capacitive semicircle is observed in the Nyquist plots in the high- and medium-frequency range, and for some tested mixtures, an inductive loop is observed in the low-frequency range. The irregular shape of the semicircle is an indicator of the impedance frequency dispersion at the interface and a characteristic of the inhomogeneity of the carbon steel surface [74,75]. In the presence of compound mixtures, the Bode plots show the appearance of phase angle peaks in the low- and high-frequency range. The phase angle maxima in the low-frequency range are higher and more extended compared to the blank system, while the phase angle maxima in the high-frequency range increase and shift to the higher frequency range with increasing compound mixture concentrations. All this indicates that the tested compound mixtures form an inhibitory film on the steel surface [39]. The recorded impedance spectra for the compound mixtures are like those for IOA but with a more prominent inhibitory effect, especially with the addition of 10 mg dm^{-3} and 15 mg dm^{-3} $CeCl_3$ to IOA. The influence of $CeCl_3$ on the improved inhibitory action of IOA is more pronounced when adding $CeCl_3$ in higher concentrations, specifically at 10 mg dm^{-3} and 15 mg dm^{-3}. Table 4. shows that with an increase in the concentration of $CeCl_3$ (from 5 mg dm^{-3} to 15 mg dm^{-3}) added to IOA (from 5 mg dm^{-3} to 20 mg dm^{-3}), high values of charge transfer resistances are achieved (from 893.2 Ω cm^2 to 6539.5 Ω cm^2). The mixture of 10 mg dm^{-3} $CeCl_3$ and 15 mg dm^{-3} IOA showed the highest charge transfer resistance, 6539.5 Ω cm^2, and the highest inhibitory efficiency, 94.92%. The Q_{dl} values in the presence of the compound mixtures are lower than the Q_{dl} values for the blank system and comparable or in some cases even lower than the Q_{dl} values for the system in the presence of IOA. Thus, in the presence of the compound mixtures, a protective layer is formed on the surface of the carbon steel that has better protective properties compared to the system in the presence of a single IOA. Although the synergistic effect of the compounds $CeCl_3$ and IOA ($S > 1$) was proven only in the case of the lowest concentrations of the compounds, 5 mg dm^{-3} and 5 mg dm^{-3}, at higher concentrations of $CeCl_3$ (10 mg dm^{-3} and 15 mg dm^{-3}), the effect of the compound mixtures in corrosion protection is clearly enhanced. The addition of $CeCl_3$ at concentrations of 10 mg dm^{-3} and 15 mg dm^{-3} to IOA caused higher inhibitory efficacy (from 67.77% to 94.92%) compared to the single IOA (from 3.43% to 93.61%).

3.3. Carbon Steel Surface Morphology

Most methods used for testing the effect of $CeCl_3$ addition to IOA on inhibitory activity showed the best effect for the mixture of 15 mg dm^{-3} $CeCl_3$ and 15 mg dm^{-3}

IOA. Therefore, this mixture of compounds was chosen for surface analysis, as well as the concentrations of the individual compounds. Figure 11 shows SEM images of the surfaces of AISI 1018 carbon steel after 72 h of exposure to a CO_2-saturated brine solution in the presence of individual compounds of 15 mg dm^{-3} $CeCl_3$ (a), 15 mg dm^{-3} IOA (b) and their mixtures (c). The corresponding EDS mapping analysis of elements on the carbon steel surface is shown in Figure 12.

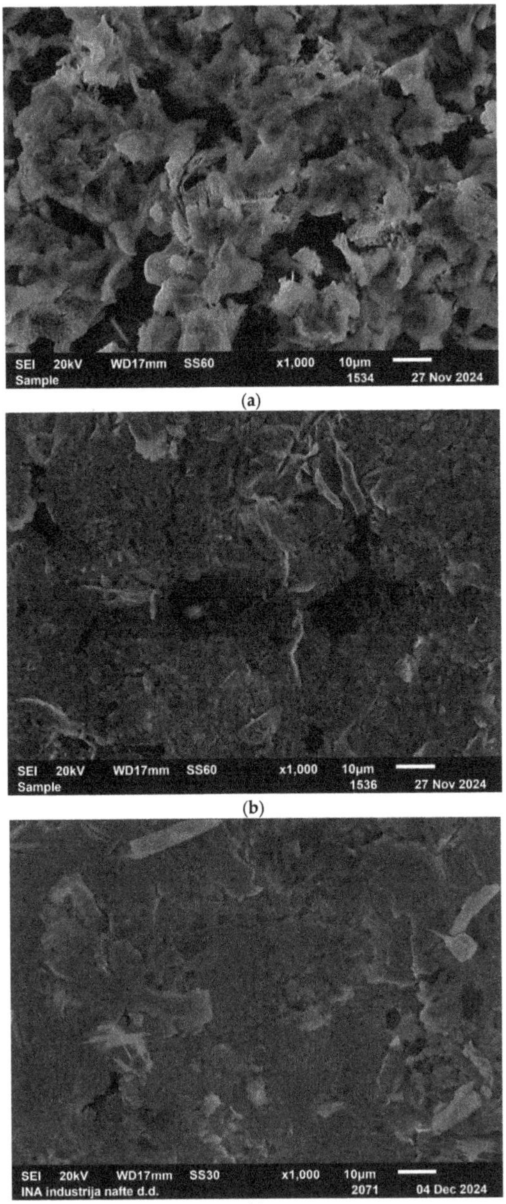

(c)

Figure 11. SEM images of carbon steel AISI 1018 surfaces after 72 h of exposure to a chloride–carbonate solution saturated with CO_2 in the presence of (**a**) 15 mg dm^{-3} $CeCl_3$, (**b**) 15 mg dm^{-3} IOA and (**c**) the mixture of 15 mg dm^{-3} $CeCl_3$ and 15 mg dm^{-3} IOA.

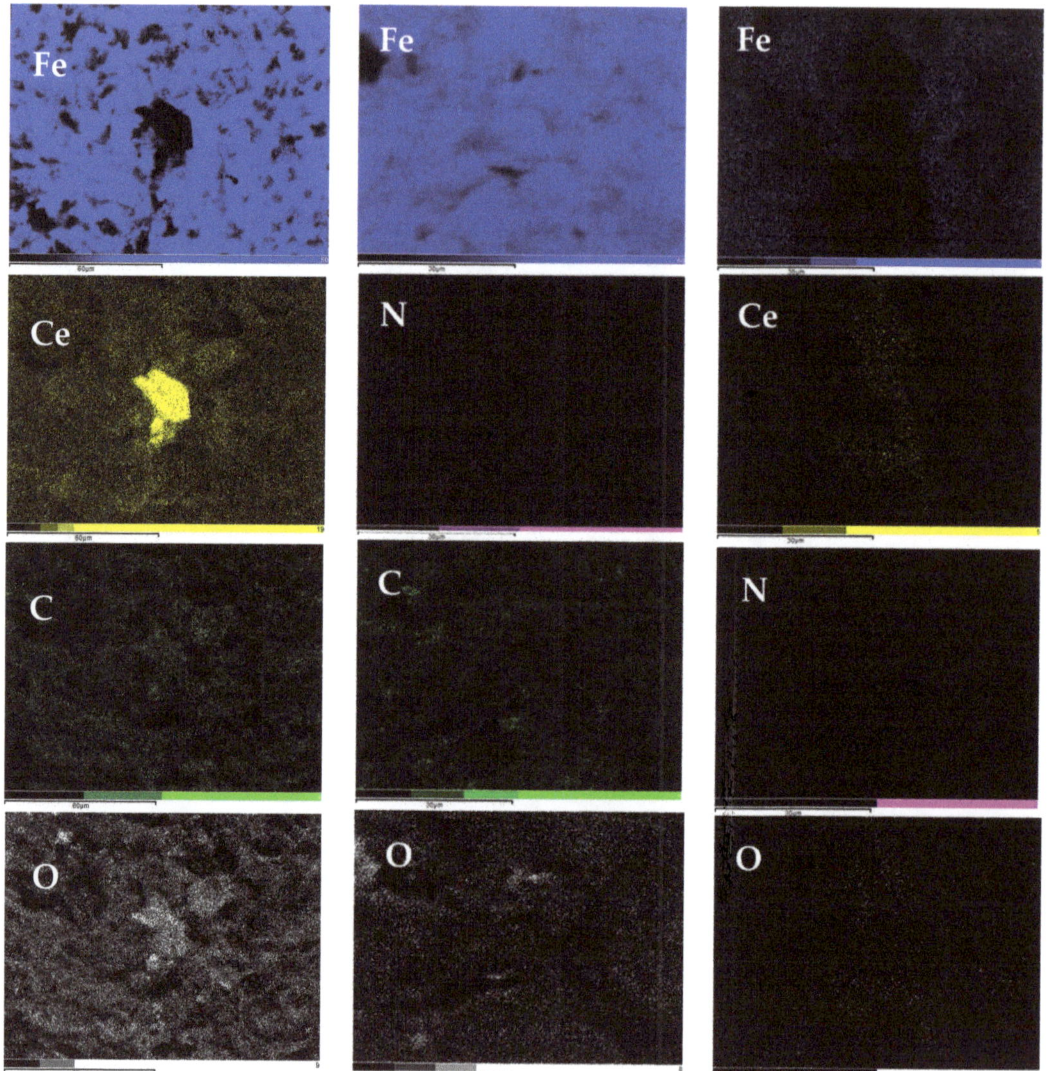

(**a**) (**b**) (**c**)

Figure 12. EDS mapping analysis of elements on carbon steel AISI 1018 surfaces after 72 h of exposure to a chloride–carbonate solution saturated with CO_2 in the presence of (**a**) 15 mg dm^{-3} $CeCl_3$ (**b**) 15 mg dm^{-3} IOA and (**c**) the mixture of 15 mg dm^{-3} $CeCl_3$ and 15 mg dm^{-3} IOA.

A SEM image of the carbon steel surface after 72 h of exposure to the synthetic brine solution saturated with CO_2 in the presence of 15 mg dm^{-3} $CeCl_3$ (Figure 11a) shows a porous and non-compact surface film. Such a surface is in accordance with the results obtained via the mass loss method (where the exposure to the tested electrolyte also lasts 72 h), whereby in the presence of $CeCl_3$, an inhibitory efficiency of 64.24% to 66.88% is achieved. EDS analysis and element mapping (Figure 12a) suggest that the formed thin film is a Ce-O-C compound, which according to XRD analysis (Figure 13), corresponds to cerium carbonate, $Ce_2(CO_3)_3$. The Raman spectrum for the system in the presence of

15 mg dm^{-3} $CeCl_3$ (Figure 14a) suggests the existence of a symmetric stretching mode of the Ce–O vibrational unit at 454 cm^{-1} [76].

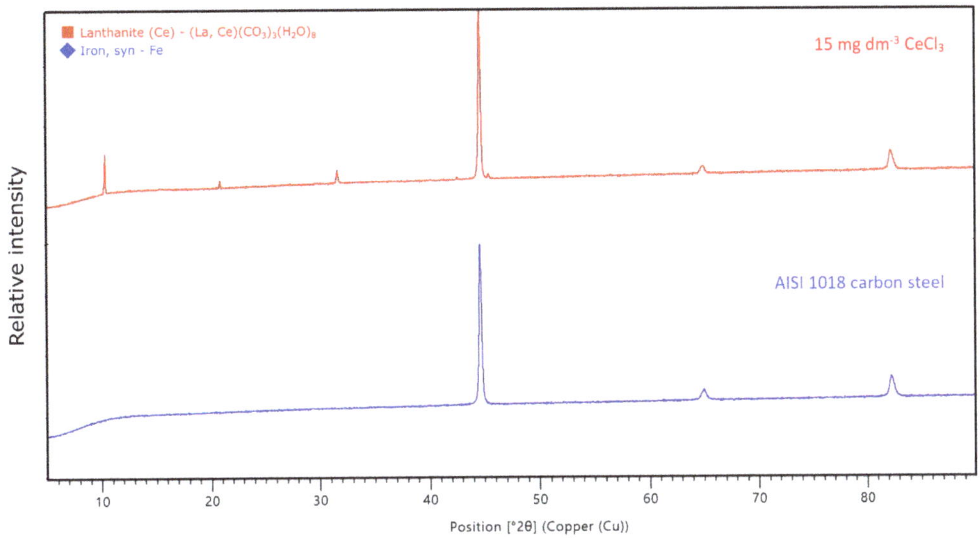

Figure 13. X-ray diffraction pattern of AISI 1018 carbon steel surface before and after exposure for 72 h to the synthetic brine solution saturated with CO_2 and with the addition of 15 mg dm^{-3} $CeCl_3$.

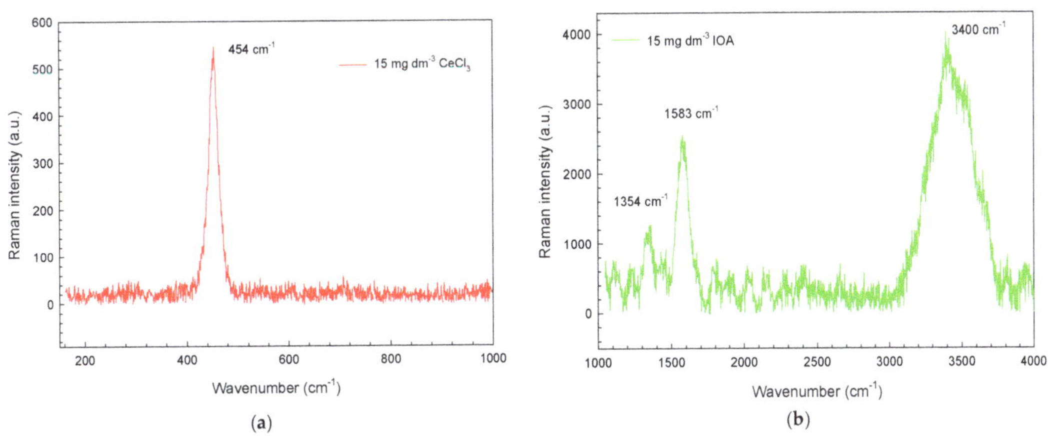

Figure 14. Raman spectra recorded on carbon steel AISI 1018 surfaces after 72 h of exposure to a chloride–carbonate solution saturated with CO_2 in the presence of (**a**) 15 mg dm^{-3} $CeCl_3$ and (**b**) 15 mg dm^{-3} IOA.

The formation of rare earth carbonates in CO_2 aqueous systems, with available CO_3^{2-} ions, is thermodynamically favorable. The mechanism of formation corresponds to the process of capturing dissolved CO_2 near the metal surface, thereby reducing the corrosiveness of the medium [77–79]. The formation of cerium carbonate, $Ce_2(CO_3)_3$, can be explained using the following reactions:

$$CeCl_{3\,(s)} \rightarrow Ce^{3+}_{(aq)} + 3Cl^{-}_{(aq)} \quad (16)$$

$$2Ce^{3+}_{(aq)} + 3CO^{2-}_{3\,(aq)} \rightarrow Ce_2(CO_3)_{3\,(s)} \quad (17)$$

The carbonate anion comes from the dissociation of carbonic acid formed according to Equation (5). The dissociation of diprotic carbonic acid occurs in two steps. The first step (Equation (6)) is fast, while the second step (Equation (7)) is slower and occurs at approximately neutral pH. Therefore, it is assumed that the precipitation of cerium carbonate is analogous to the precipitation of iron carbonate in an acidic medium according to the following reactions [77,80]:

$$2Ce^{3+}_{(aq)} + 3HCO^{-}_{3\ (aq)} \rightarrow Ce_2(HCO_3)_{3\ (s)} \quad (18)$$

$$Ce_2(HCO_3)_{3\ (s)} \rightarrow Ce_2(CO_3)_{3\ (s)} + 3H^{+}_{(aq)}. \quad (19)$$

A SEM image of the carbon steel surface after 72 h of exposure to the synthetic brine solution saturated with CO_2 in the presence of 15 mg dm^{-3} IOA (Figure 11b) shows a denser film compared to the surface layer in the presence of $CeCl_3$. However, voids are observed, indicating that the layer does not have fully protective corrosion properties. The adsorption of imidazoline on the steel surface can be physical or chemical. In its molecular structure (Figure 2a), imidazoline has a heterocyclic ring with a double covalent bond (π-delocalized electrons) as well as free electron pairs on nitrogen atoms. These are also active sites through which IOA can be attached to the steel surface. In the case of chemisorption, it achieves a coordinative type of covalent bond in which it fills the empty d-orbitals of iron with its electrons, while in physical adsorption, an electrostatic interaction is achieved between the negatively charged part of the IOA molecule (π-electrons or free electron pairs on nitrogen) and the positively charged surface of the metal. Also, the hydrophobic hydrocarbon chain on the ring of the IOA molecule hinders the interaction of water molecules with the steel surface [40,81]. Adsorption through the active sites of the IOA molecule is confirmed by EDS element mapping (Figure 12b) and recorded Raman spectrum (Figure 14b). The Raman spectrum for pure IOA is shown in Figure 14b. The band at 1354 cm^{-1} corresponds to C-N stretching, while the band at 1583 cm^{-1} corresponds to N=C stretching in the heteroatomic ring [82,83]. The broad peak around 3400 cm^{-1} represents the asymmetric O-H stretching region for water [84].

The SEM image of the steel surface after exposure to a mixture of $CeCl_3$ and IOA (Figure 11c) shows the existence of a dense and compact surface film. Low corrosion rates and high inhibitory effectiveness determined via the mass loss method after 72 h of exposure to a mixture of 15 mg dm^{-3} IOA and 15 mg dm^{-3} $CeCl_3$ prove that the film created in that time period has good protective properties. The Raman spectrum of the steel surface after exposure to the mixture of compounds (Figure 15) shows the existence of bands at around 1200 cm^{-1} and 500 cm^{-1}, which correspond to C-N stretching and Ce-O stretching, respectively [76,82,83]. The small shift in stretching modes compared to individual components is a consequence of the interaction of components adsorbed on the electrode surface. The improved/synergistic effect of the mixture of compounds and carbon steel surface layer after exposure to mixture can be explained by the pore plugging effect of cerium carbonate [85,86]. Cerium carbonate fills the pores of the IOA adsorbed film and makes the surface film more compact and with better protective properties.

Surface analyses showed the existence of protective films on the surface of carbon steel after 72 h of exposure to the system in the presence of individual components, $CeCl_3$ and IOA, as well as their mixtures. However, the protective properties of the film formed in the presence of compound mixtures are much better, as confirmed by the results obtained via the mass loss method. Electrochemical tests show high inhibitory efficiencies of individual compounds; however, since they were carried out after 24 h of exposure to the tested electrolytes, it is possible that their efficiency decreases with time. The mass loss test lasted for 72 h, after which the achieved inhibitory efficacies of the tested individual compounds, $CeCl_3$ and IOA, were significantly lower compared to the values obtained via electrochemical methods. The protective properties of the inhibitory layers of individual compounds decrease with time, possibly due to the desorption of the adsorbed inhibitor. Furthermore, the mixture of compounds showed synergistic action and better corrosion protection (lower corrosion rates and higher inhibitory efficacies) than the individual compounds. Cerium, as well as other rare earth elements, in combination with other compounds/coatings, is known for its so-called self-healing properties that depend on pH value and time [87,88]. Obviously, with longer exposure time, the porosity of the primary inhibitor film formed by IOA decreases, since the pores are filled with cerium carbonate. Such a layer is more compact and protects better against corrosion, but it is probably formed over a longer period (certainly more than 24 h).

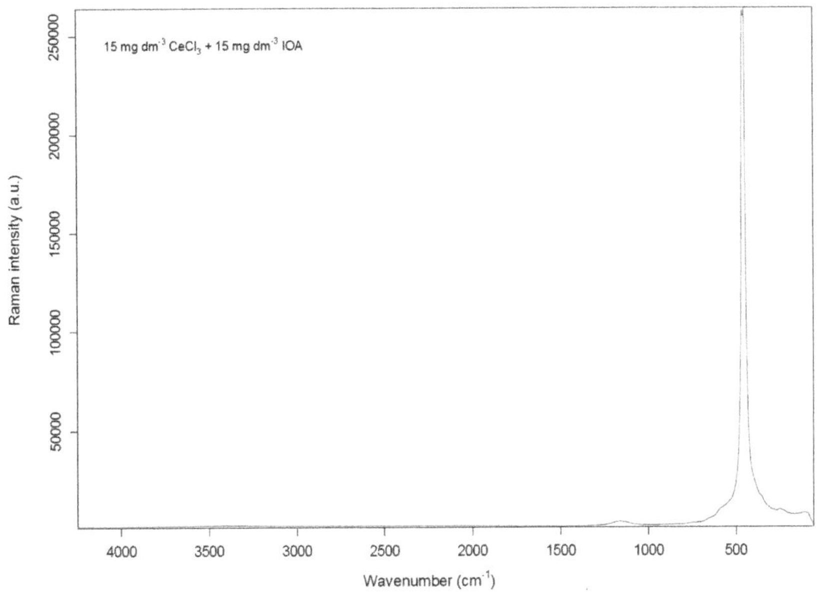

Figure 15. Raman spectra recorded on carbon steel AISI 1018 surfaces after 72 h of exposure to a chloride–carbonate solution saturated with CO_2 in the presence of the mixture of 15 mg dm^{-3} $CeCl_3$ and 15 mg dm^{-3} IOA.

4. Conclusions

Based on the test results, for the sake of simplicity of this review, Table 5 shows the concentrations of individual components as well as the concentrations of the compound mixtures, where they achieve the lowest corrosion rates, v_{corr}, and the highest inhibitory efficacy, *IE*, (divided by test methods).

Table 5. Selected concentrations of individual components and their mixtures with the corresponding lowest corrosion rates and highest inhibitory efficiencies proven via individual testing methods.

	Mass Loss Method			Linear Polarization			Potentiodynamic Polarization			EIS *	
System	γ (mg dm^{-3})	v_{corr} (mm year^{-1})	*IE* (%)	γ (mg dm^{-3})	v_{corr} (mm year^{-1})	*IE* (%)	γ (mg dm^{-3})	v_{corr} (mm year^{-1})	*IE* (%)	γ (mg dm^{-3})	*IE* (%)
SBS	0	1.57	/	0	1.27	/	0	1.45	/	0	/
IOA	20	0.17	89.17	20	0.08	92.89	15	0.08	94.61	20	93.61
$CeCl_3$	10	0.52	66.88	20	0.06	95.65	15	0.05	96.27	20	90.96
$CeCl_3$ + IOA	15 + 15	0.05	96.82	15 + 15	0.03	97.94	15 + 15	0.03	97.78	10 + 15	94.92

* The EIS method does not determine the corrosion rate.

- Mixtures of tested compounds, IOA and $CeCl_3$, showed mainly enhanced activity in corrosion protection compared to individual activity;
- Most of test methods showed that the mixture of compounds of 15 mg dm^{-3} IOA and 15 mg dm^{-3} CeCl3 achieved the lowest corrosion rate (0.03 mm $year^{-1}$–0.05 mm $year^{-1}$) and the highest inhibitory efficiency (94.92%–97.94%);
- IOA can be categorized as an anodic corrosion inhibitor, while $CeCl_3$ and the mixture of $CeCl_3$ and IOA act as mixed corrosion inhibitors with a predominant influence on the anodic process;
- The inhibitory effect of individual components and their mixtures is realized by the formation of a protective film on the steel surface: IOA by its physical/chemical adsorption on the steel surface and $CeCl_3$ by the formation of a protective layer of cerium carbonate, while the mixture of IOA and $CeCl_3$ creates a porous layer of adsorbed IOA, with cerium carbonate filling these pores;
- The results obtained via the mass loss method and surface analysis can be considered more reliable compared to electrochemical tests since they provide information about the effectiveness and synergy of the compounds over a longer period (72 h) compared to electrochemical tests (24 h);
- According to results obtained via the mass loss method and surface analysis, the effectiveness of individual components decreases with time, but the effectiveness of mixtures of compounds is stable and high, and their synergistic action was determined;
- To gain a better insight into the inhibitory effect of the tested compounds and their synergy, electrochemical tests will be conducted in the future over a longer period; measurements will be carried out at different temperatures, and the nature of the adsorption of individual compounds and their mixtures will be determined.

Author Contributions: Conceptualization, T.B., G.B., K.Ž. and H.O.Ć.; Methodology, G.B.; Formal analysis, T.B., G.B., K.Ž. and H.O.Ć.; Investigation, T.B.; Resources, G.B. and T.B.; Writing—original draft preparation, T.B. and G.B.; Writing—review and editing, T.B., G.B., K.Ž. and H.O.Ć.; Visualization, T.B. and G.B.; Supervision, G.B., K.Ž. and H.O.Ć.; Project administration, G.B. and K.Ž.; Funding acquisition, K.Ž. All authors have read and agreed to the published version of the manuscript.

Funding: This research received no external funding.

Institutional Review Board Statement: Not applicable.

Informed Consent Statement: Not applicable.

Data Availability Statement: Data is contained within the article.

Conflicts of Interest: Author Tihomir Borko was employed by INA, d.d. The remaining authors declare that the research was conducted in the absence of any commercial or financial relationships that could be construed as a potential conflict of interest.

References

1. Sharifzadeh, M.; Triulzi, G.; Magee, C.L. Quantification of technological progress in greenhouse gas (GHG) capture and mitigation using patent data. *Energy Environ. Sci.* **2019**, *12*, 2789–2805. [CrossRef]
2. Gu, S.; Li, Y.; Teng, L.; Wang, C.; Hu, Q.; Zhang, D.; Ye, X.; Wang, J.; Iglauer, S. An experimental study on the flow characteristics during the leakage of high pressure CO_2 pipelines. *Process Saf. Environ. Protect.* **2019**, *125*, 92–101. [CrossRef]
3. Jiang, K.; Ashworth, P. The development of Carbon Capture Utilization and Storage (CCUS) research in China: A bibliometric perspective. *Renew. Sustain. Energy Rev.* **2021**, *138*, 110521. [CrossRef]
4. Wang, C.; Li, Y.; Teng, L.; Gu, S.; Hu, Q.; Zhang, D.; Ye, X.; Wang, J. Experimental study on dispersion behavior during the leakage of high pressure CO_2 pipelines. *Exp. Therm. Fluid Sci.* **2019**, *105*, 77–84. [CrossRef]
5. Ajayi, T.; Gomes, J.S.; Bera, A. A review of CO_2 storage in geological formations emphasizing modeling, monitoring and capacity estimation approaches. *Petrol. Sci.* **2019**, *16*, 1028–1063. [CrossRef]

6. Xu, Z.X.; Li, S.Y.; Li, B.F.; Chen, D.Q.; Liu, Z.Y.; Li, Z.M. A review of development methods and EOR technologies for carbonate reservoirs. *Petrol. Sci.* **2020**, *17*, 990–1013. [CrossRef]
7. Hua, Y.; Shamsa, A.; Barker, R.; Neville, A. Protectiveness, morphology and composition of corrosion products formed on carbon steel in the presence of Cl^-, Ca^{2+} and Mg^{2+} in high pressure CO_2 environments. *Appl. Surf. Sci.* **2018**, *455*, 667–682. [CrossRef]
8. Li, Y.; Liu, X.; Wang, C.; Hu, Q.; Wang, J.; Ma, H.; Zhang, N. Research progress on corrosion behavior of gaseous CO_2 transportation pipelines containing impurities. *Acta Metall. Sin.* **2020**, *57*, 283–294. [CrossRef]
9. Nesic, S. Effects of multiphase flow on internal CO_2 corrosion of mild steel pipelines. *Energy Fuel.* **2012**, *26*, 4098–4111. [CrossRef]
10. Wang, Z.M.; Song, G.L.; Zhang, J. Corrosion Control in CO_2 Enhanced Oil Recovery From a Perspective of Multiphase Fluids. *Front. Mater.* **2019**, *6*, 272. [CrossRef]
11. Yoon-Seok, C.; Nesic, S. Determining the corrosive potential of CO_2 transport pipeline in high pCO_2–water environments. *Int. J. Greenh. Gas Control* **2011**, *5*, 788–797.
12. Xiang, Y. Corrosion issues of carbon capture, utilization, and storage. *Mater. Perform.* **2018**, *57*, 32–35.
13. Cao, S.; He, F.; Gao, J. Corrosion problems in the oil country tubular goods and their mitigation—A review. *Anti-Corros. Method M.* **2017**, *64*, 465–478. [CrossRef]
14. Obot, I.B.; Onyeachu, I.B.; Umoren, S.A.; Quraishi, M.A.; Sorour, A.A.; Chen, T.; Aljeaban, N.; Wang, Q. High temperature sweet corrosion and inhibition in the oil and gas industry: Progress, challenges and future perspectives. *J. Petrol. Sci. Eng.* **2020**, *185*, 106469. [CrossRef]
15. Onyeachu, I.B.; Chauhan, D.S.; Quraishi, M.A.; Obot, I.B. Influence of hydrodynamic condition on 1,3,5-tris(4-methoxyphenyl)-1,3,5-triazinane as a novel corrosion inhibitor formulation for oil and gas industry. *Corros. Eng. Sci. Technol.* **2021**, *56*, 154–161. [CrossRef]
16. Kermani, M.B.; Harrop, D. The impact of corrosion on oil and gas industry. *SPE Prod. Oper.* **1996**, *11*, 186–190. [CrossRef]
17. Kermani, B.M.; Morshed, A. Carbon dioxide corrosion in oil and gas production: A compendium. *Corrosion* **2003**, *59*, 659–683. [CrossRef]
18. Paolinelli, L.D.; Perez, T.; Simison, S.N. The effect of pre-corrosion and steel microstructure on inhibitor performance in CO_2 corrosion. *Corros. Sci.* **2008**, *50*, 2456–2464. [CrossRef]
19. Sastri, V.S. *Green Corrosion Inhibitors: Theory and Practice*; John Wiley & Sons, Inc.: Hoboken, NJ, USA, 2011.
20. Darling, D.; Rakshpal, R. Green Chemistry Applied to Corrosion and Scale Inhibitors. In Proceedings of the CORROSION, San Diego, CA, USA, 22–27 March 1998; p. 98207.
21. Progress in Corrosion—The First 50 Years of the EFC, Edited by P. McIntyre & J. Vogelsang, Published for the European Federation of Corrosion by Maney Publishing on Behalf of the Institute of Materials, Minerals & Mining. 2009. Available online: http://ndl.ethernet.edu.et/bitstream/123456789/64636/1/1371.pdf (accessed on 16 October 2024.).
22. de Damborenea, J.; Conde, A.; Arenas, M.A. Corrosion inhibition with rare earth metal compounds in aqueous solutions. In *Rare Earth-Based Corrosion Inhibitors*; Woodhead Publishing: Cambridge, UK, 2014; pp. 84–116. [CrossRef]
23. Ahmed, M.A.; Amin, S.; Mohamed, A.A. Current and emerging trends of inorganic, organic and eco-friendly corrosion inhibitors. *RSC Adv.* **2024**, *14*, 31877–31920. [CrossRef]
24. Freiner, W.W. Review of Green Chemistry Corrosion Inhibitors for Aqueous Systems. In Proceedings of the 9th European Symposium on Corrosion Inhibitors, Ferrara, Italy, 4–8 September 2000; Universidad de Ferrara: Ferrara, Italy, 2000; Volume 1, pp. 1–37.
25. Satyanarayana Gupta, D.V. What is Green? In Proceedings of the CORROSION, San Diego, CA, USA, 22–27 March 1998; p. 206.
26. Hinton, B.R.W. Corrosion inhibition with rare earth metal salts. *J. Alloys Compd.* **1992**, *180*, 15–25. [CrossRef]
27. Mendez, J.A.; Vong, Y.M.; Bueno, J.d. Cerium and Other Rare Earth Salts as Corrosion Inhibitors—A Review. *Prot. Met. Phys. Chem. Surf.* **2022**, *58*, 801–810. [CrossRef]
28. Evans, C.H. *Biochemistry of the Lanthanides*; Springer Science + Business Media: New York, NY, USA, 1990.
29. Kawagoe, M.; Ishikawa, K.; Wang, S.C.; Yoshikawa, K.; Arany, S.; Zhou, X.P.; Wang, J.S.; Ueno, Y.; Koizumi, Y.; Kameda, T.; et al. Acute effects on the lung and the liver of oral administration of cerium chloride on adult, neonatal and fetal mice. *J. Trace Elem. Med. Biol.* **2008**, *22*, 59–65. [CrossRef] [PubMed]
30. Muecke, G.K.; Moller, P. The not- so rare earths. *Sci. Am.* **1988**, *258*, 72–77. [CrossRef]
31. Sastri, V.S.; Perumareddi, J.R.; Ramachandra Rao, V.; Rayudu, G.V.S.; Bünzli, J.C.G. *Modern Aspects of Rare Earths and Their Complexes*, 1st ed.; Elsevier Science: Amsterdam, The Netherlands, 2003.
32. Mohammedi, D.; Ismail, F.; Rehamnia, R.; Bensalem, R.; Savadogo, O. Corrosion behaviour of steel in the presence of rare earth salts: Synergistic effect. *Corros. Eng. Sci. Technol.* **2015**, *50*, 633–638. [CrossRef]
33. Bernal, S.; Botana, F.J.; Calvino, J.J.; Marcos, M.; Perez-Omil, J.A.; Vidal, H. Lanthanide salts as alternative corrosion inhibitors. *J. Alloys Compd.* **1995**, *225*, 638–641. [CrossRef]
34. Davó, B.; de Damborenea, J.J. Use of rare earth salts as electrochemical corrosion inhibitors for an Al–Li–Cu (8090) alloy in 3.56% NaCl. *Electrochim. Acta* **2004**, *49*, 4957–4965. [CrossRef]

35. Dastgheib, A.; Attar, M.M.; Zarebidaki, A. Evaluation of Corrosion Inhibition of Mild Steel in 3.5 wt% NaCl Solution by Cerium Nitrate. *Met. Mater. Int.* **2020**, *26*, 1634–1642. [CrossRef]
36. Chen, X.-B.; Cain, T.; Scully, J.R.; Birbilis, N. Technical Note: Experimental Survey of Corrosion Potentials for Rare Earth Metals Ce, Er, Gd, La, and Nd as a Function of pH and Chloride Concentration. *Corrosion* **2013**, *70*, 323–328. [CrossRef]
37. Arenas, M.A.; Conde, A.; de Damborenea, J.J. Cerium: A suitable green corrosion inhibitor for Tinplate. *Corros. Sci.* **2002**, *44*, 511–520. [CrossRef]
38. Okafor, P.C.; Liu, C.; Liu, X.; Zheng, Y.; Wang, F.; Lin, C.Y. Corrosion inhibition and adsorption behavior of imidazoline salt on N80 carbon steel in CO_2-saturated solutions and its synergism with thiourea. *J. Solid State Electrochem.* **2010**, *14*, 1367–1376. [CrossRef]
39. Heydari, M.; Javidi, M. Corrosion inhibition and adsorption behaviour of an amido-imidazoline derivative on API 5L X52 steel in CO_2-saturated solution and synergistic effect of iodide ions. *Corros. Sci.* **2012**, *61*, 148–155. [CrossRef]
40. Munis, A.; Zhao, T.; Zheng, M.; Rehman, A.U.; Wang, F. A newly synthesized green corrosion inhibitor imidazoline derivative for carbon steel in 7.5% NH_4Cl solution. *Sustain. Chem. Pharm.* **2020**, *16*, 1000258. [CrossRef]
41. Zhao, J.; Duan, H.; Jiang, R. Synergistic corrosion inhibition effect of quinoline quaternary ammonium salt and Gemini surfactant in H_2S and CO_2 saturated brine solution. *Corros. Sci.* **2015**, *91*, 108–119. [CrossRef]
42. Zhang, D.; Wu, H.; Gao, L. Synergistic inhibition effect of l-phenylalanine and rare earth Ce(IV) ion on the corrosion of copper in hydrochloric acid solution. *Mater. Chem. Phys.* **2012**, *133*, 981–986. [CrossRef]
43. Marcelin, S.; Pébère, N. Synergistic effect between 8-hydroxyquinoline and benzotriazole for the corrosion protection of 2024 aluminium alloy: A local electrochemical impedance approach. *Corros. Sci.* **2015**, *101*, 66–74. [CrossRef]
44. Khamis, A.; Saleh, M.M.; Awad, M.I. Synergistic inhibitor effect of cetylpyridinium chloride and other halides on the corrosion of mild steel in 0.5 M H_2SO_4. *Corros. Sci.* **2013**, *66*, 343–349. [CrossRef]
45. Wang, Y.; Northwood, D.O. Effects of O_2 and H_2 on the corrosion of SS316L metallic bipolar plate materials in simulated anode and cathode environments of PEM fuel cells. *Electrochim. Acta* **2007**, *52*, 6793–6798. [CrossRef]
46. Bahrami, M.J.; Hosseini, S.M.A.; Pilvar, P. Experimental and theoretical investigation of organic compounds as inhibitors for mild steel corrosion in sulfuric acid medium. *Corros. Sci.* **2010**, *52*, 2793–2803. [CrossRef]
47. Elgaddafi, R.; Naidu, A.; Ahmed, R.; Shah, S.; Hassani, S.; Osisanya, S.O.; Saasen, A. Modeling and experimental study of CO_2 corrosion on carbon steel at elevated pressure and temperature. *J. Nat. Gas Sci. Eng.* **2015**, *27*, 1620–1629. [CrossRef]
48. Tan, J.; Guo, L.; Yang, H.; Zhang, F.; El Bakri, Y. Synergistic effect of potassium iodide and sodium dodecyl sulfonate on the corrosion inhibition of carbon steel in HCl medium: A combined experimental and theoretical investigation. *RSC Adv.* **2020**, *10*, 15163–15170. [CrossRef]
49. Li, X.; Deng, S.; Du, G.; Xie, X. Synergistic inhibition effect of walnut green husk extract and sodium lignosulfonate on the corrosion of cold rolled steel in phosphoric acid solution. *J. Taiwan Inst. Chem. Eng.* **2020**, *114*, 263–283. [CrossRef]
50. Wang, Z.; Wang, T.; Zhu, J.; Wei, L.; Shen, Y.; Li, N.; Hu, J. Synergistic effect and mechanism of copper corrosion inhibition using cinnamaldehyde and vanillin in HCl solution: An experimental and theoretical approach. *Colloids Surf. A Physicochem. Eng. Asp.* **2019**, *563*, 246–254. [CrossRef]
51. Pustaj, G.; Kapor, F.; Kvasnička, P. Inhibition properties and adsorption behavior of olive leaf extract on N80 carbon steel in CO_2-saturated brine solution. *Int. J. Mater. Res.* **2014**, *105*, 992–998. [CrossRef]
52. Žbulj, K.; Bilić, G.; Simon, K.; Hrnčević, L. Lady's Mantle Flower as a Biodegradable Plant-Based Corrosion Inhibitor for CO_2 Carbon Steel Corrosion. *Coatings* **2024**, *14*, 671. [CrossRef]
53. Tran, T.; Brown, B.N.; Nesic, S. Corrosion of Mild Steel in an Aqueous CO_2 Environment—Basic Electrochemical Mechanisms Revisited. In Proceedings of the NACE—International Corrosion Conference Series 2015, Conference & Expo, Dallas, TX, USA, 15–19 March 2015. Paper No. 5671.
54. Kahyarian, A.; Nesic, S. On the mechanism of carbon dioxide corrosion of mild steel: Experimental investigation and mathematical modeling at elevated pressures and non-ideal solutions. *Corros. Sci.* **2020**, *173*, 108719. [CrossRef]
55. El-Katori, E.E.; Fouda, A.S.; Mohamed, R.R. Synergistic corrosion inhibition activity of the *Chicoriumintybus* extract and iodide ions for mild steel in acidic media. *J. Chil. Chem. Soc.* **2020**, *65*, 4672–4681. [CrossRef]
56. Rahayu, D.U.C.; Cahyani, S.; Abdullah, I.; Nurani, D.A.; Krisnandi, Y.K. Microwave-assisted synthesis of organic corrosion inhibitor based imidazoline-stearic. *IOP Conf. Ser. Mater. Sci. Eng.* **2019**, *902*, 012019. [CrossRef]
57. Porcayo-Calderon, J.; Canto, J.; Martinez de la Escalera, L.M.; Neri, A. Sweet Corrosion Inhibition by CO_2 Capture. *Molecules* **2022**, *27*, 5209. [CrossRef]
58. Martinez de la Escalera, D.M.; Ramos-Hernandez, J.J.; Porcayo-Palafox, E.; Porcayo-Calderon, J.; Gonzalez-Rodriguez, J.G.; Martinez-Gomez, L. Effect of Nd^{3+} Ion Concentration on the Corrosion Resistance of API X70 Steel in Chloride-Rich Environments. *Adv. Mater. Sci. Eng.* **2018**, *2018*, 9328317. [CrossRef]

59. Porcayo-Calderon, J.; Ramos-Hernandez, J.J.; Porcayo-Palafox, E.; Martinez de la Escalera, L.M.; Canto, J.; Gonzalez-Rodriguez, J.G.; Martinez-Gomez, L. Sustainable Development of Corrosion Inhibitors from Electronic Scrap: Synthesis and Electrochemical Performance. *Adv. Mater. Sci. Eng.* **2019**, *2019*, 6753658. [CrossRef]
60. Zhang, G.; Chen, C.; Lu, M.; Chai, C.; Wu, Y. Evaluation of inhibition efficiency of an imidazoline derivative in CO_2-containing aqueous solution. *Mater. Chem. Phys.* **2007**, *105*, 331–340. [CrossRef]
61. Lopez, D.A.; Simison, S.N.; de Sanchez, S.R. The influence of steel microstructure on CO_2 corrosion. EIS studies on the inhibition efficiency of benzimidazole. *Electrochem. Acta* **2003**, *48*, 845–854. [CrossRef]
62. Pustaj, G.; Kapor, F.; Jakovljević, S. Carbon dioxide corrosion of carbon steel and corrosion inhibition by natural olive leaf extract: Die Kohlendioxid-Korrosion von Kohlenstoffstahl und die Korrosionshemmung durch natürlichen Olivenblattextrakt. *Mater. Werkst.* **2017**, *48*, 122–138. [CrossRef]
63. Pustaj, G.; Kapor, F.; Veinović, Ž. Olive Leaf Extract as a Corrosion Inhibitor of Carbon Steel in CO_2-Saturated Chloride–Carbonate Solution. *Int. J. Electrochem. Sci.* **2016**, *11*, 7811–7829. [CrossRef]
64. Herrera Hernández, H.; González Díaz, F.; Fajardo SanMiguel, G.; Velázquez Altamirano, J.C.; González Morán, C.O.; Morales Hernández, J. Electrochemical impedance spectroscopy as a practical tool for monitoring the carbonation process on reinforced concrete structures. *Arab. J. Sci. Eng.* **2019**, *44*, 10087–10103. [CrossRef]
65. Herrera Hernández, H.; Ruiz Reynoso, A.M.; Trinidad González, J.C.; Gonzales Moran, C.O.; Miranda Hernandez, J.G.; Mandujano Ruiz, A.; Morales Hernandez, J.; Orozco Cruz, R. Electrochemical Impedance Spectroscopy (EIS): A Review Study of Basic Aspects of the Corrosion Mechanism Applied to Steels. In *Electrochemical Impedance Spectroscopy*; IntechOpen: London, UK, 2020. [CrossRef]
66. Desimone, M.P.; Grundmeier, G.; Gordillo, G.; Simison, S.N. Amphiphilic amido-amine as an effective corrosion inhibitor for mild steel exposed to CO_2 saturated solution: Polarization, EIS and PM-IRRAS studies. *Electrochim. Acta* **2011**, *56*, 2990–2998. [CrossRef]
67. El-Azazy, M.; Mart, M.; Annus, P. (Eds.) *Electrochemical Impedance Spectroscopy*; IntechOpen: London, UK, 2020. [CrossRef]
68. Sun, J.B.; Zhang, G.A.; Liu, W.; Lu, M.X. The formation mechanism of corrosion scale and electrochemical characteristic of low alloy steel in carbon dioxide-saturated solution. *Corros. Sci.* **2012**, *57*, 131–138. [CrossRef]
69. Nam, N.D.; Bui, Q.V.; Mathesh, M.; Tan, M.Y.J.; Forsyth, M. A study of 4-carboxyphenylboronic acid as a corrosion inhibitor for steel in carbon dioxide containing environments. *Corros. Sci.* **2013**, *76*, 257–266. [CrossRef]
70. Singh, A.K.; Quraishi, M.A. Effect of Cefazolin on the corrosion of mild steel in HCl solution. *Corros. Sci.* **2010**, *52*, 152–160. [CrossRef]
71. Jia, W. A study on the impedance responses of inhibitor desorption. *Chin. J. Ocean. Limnol.* **1998**, *16*, 54–59. [CrossRef]
72. Okafor, P.C.; Liu, X.; Zheng, Y. Corrosion inhibition of mild steel by ethylamino imidazoline derivative in CO_2-saturated solution. *Corros. Sci.* **2009**, *51*, 761–768. [CrossRef]
73. Videm, K. The Anodic Behavior of Iron and Steel in Aqueous Solutions with CO_2, HCO_3^-, CO_3^{2-} and CI^-. In Proceedings of the CORROSION, Orlando, FL, USA, 26–31 March 2000; p. 39.
74. Li, L.J.; Zhang, X.P.; Lei, J.L.; Pan, F.S. Osmanthus Fragran Leaves Extract as Corrosion Inhibitor for Carbon Steel in Hydrochloric Acid Solution. *Asian J. Chem.* **2012**, *24*, 1649–1653.
75. Bentiss, F.; Lebrini, M.; Vezin, H.; Chai, F.; Traisnel, M.; Lagrené, M. Enhanced corrosion resistance of carbon steel in normal sulfuric acid medium by some macrocyclic polyether compounds containing a 1,3,4-thiadiazole moiety: AC impedance and computational studies. *Corros. Sci.* **2009**, *51*, 2165–5173. [CrossRef]
76. Kim, N.W.; Lee, D.K.; Yu, H. Selective shape control of cerium oxide nanocrystals for photocatalytic and chemical sensing effect. *RSC Adv* **2019**, *9*, 13829–13837. [CrossRef]
77. Anadebe, V.C.; Chukwuike, V.I.; Ramanathan, S.; Barik, R.C. Cerium-based metal organic framework (Ce-MOF) as corrosion inhibitor for API 5L X65 steel in CO_2- saturated brine solution: XPS, DFT/MD-simulation, and machine learning model prediction. *Process Saf. Environ. Prot.* **2022**, *168*, 499–512. [CrossRef]
78. Hassas, B.V.; Rezaee, M.; Pisupati, S.V. Precipitation of rare earth elements from acid mine drainage by CO_2 mineralization process. *Chem. Eng. J.* **2020**, *399*, 125716. [CrossRef]
79. Pei, S.-L.; Pan, S.-Y.; Li, Y.-M.; Chiang, P.-C. Environmental benefit assessment for the carbonation process of petroleum coke fly ash in a rotating packed bed. *Environ. Sci. Technol.* **2017**, *51*, 10674–10681. [CrossRef]
80. Barker, R.; Burkle, D.; Charpentier, T.; Thompson, H.; Neville, A. A review of iron carbonate ($FeCO_3$) formation in the oil and gas industry. *Corros. Sci.* **2018**, *142*, 312–341. [CrossRef]
81. Tyagi, R.; Tyagi, V.K.; Pandey, S.K. Imidazoline and Its Derivatives: An Overview. *J. Oleo Sci.* **2007**, *56*, 211–222. [CrossRef]
82. Shimpi, M.R.; Childs, S.L.; Bostrom, D.; Velaga, S.P. New cocrystals of ezetimibe with l-proline and imidazole. *CrystEngComm* **2014**, *16*, 8984–8993. [CrossRef]
83. Raman Band Correlation Table. Available online: https://www.chem.uci.edu/~dmitryf/manuals/Raman%20correlations.pdf (accessed on 10 December 2024).

84. Durickovic, I. Using Raman Spectroscopy for Characterization of Aqueous Media and Quantification of Species in Aqueous Solution. In *Applications of Molecular Spectroscopy to Current Research in the Chemical and Biological Sciences*; Stauffer, M.T., Ed.; InTech: London, UK, 2016. [CrossRef]
85. Penga, Y.; Hughes, A.E.; Mardel, J.I.; Deacon, G.B.; Junk, P.C.; Catubig, R.A.; Forsyth, M.; Hinton, B.R.W.; Somers, A.E. Dual function of rare earth carboxylate compounds on the barrier properties and active corrosion inhibition of epoxy coatings on mild steel. *Prog. Org. Coat.* **2023**, *185*, 107870. [CrossRef]
86. Ivušić, F.; Lahodny-Šarc, O.; Otmačić Ćurković, H.; Alar, V. Synergistic inhibition of carbon steel corrosion in seawater by cerium chloride and sodium gluconate. *Corros. Sci.* **2015**, *98*, 88–97. [CrossRef]
87. Habib, S.; Fayyad, E.; Nawaz, M.; Khan, A.; Shakoor, R.A.; Kahraman, R.; Abdullah, A. Cerium Dioxide Nanoparticles as Smart Carriers for Self-Healing Coatings. *Nanomaterials* **2020**, *10*, 791. [CrossRef] [PubMed]
88. Guergova, D.; Stoyanova, E.; Stoychev, D.; Avramova, I.; Stefanov, P. Self-healing effect of ceria electrodeposited thin films on stainless steel in aggressive 0.5 mol/L NaCl aqueous solution. *J. Rare Earths* **2015**, *33*, 1212–1227. [CrossRef]

Disclaimer/Publisher's Note: The statements, opinions and data contained in all publications are solely those of the individual author(s) and contributor(s) and not of MDPI and/or the editor(s). MDPI and/or the editor(s) disclaim responsibility for any injury to people or property resulting from any ideas, methods, instructions or products referred to in the content.

coatings

MDPI

Article

The Damage Evolution of a Cr_2O_3-TiO_2 Coating Subjected to Cyclic Impact and Corrosive Environments and the Influence of a Nickel Intermediate Layer

Huaxing Yang [1], Yang Zhao [2], Xudong Qin [1], Yixin Jin [1], Xinyang Zhao [1], Kailin Tian [1], Xiaoming Wang [2,*] and Zhao Zhang [1,*]

[1] College of Electromechanical and Automotive Engineering, Yantai University, Yantai 264000, China
[2] Army Academy of Armored Forces, Beijing 100000, China
* Correspondence: uwangxm@126.com (X.W.); trojanz@yeah.net (Z.Z.)

Abstract: Cyclic impacts in corrosive environments significantly affect the service life of ceramic coatings, greatly increasing their susceptibility to cracking and delamination. This study investigated the damage evolution behavior of Cr_2O_3-TiO_2 (CT) coatings under cyclic stress in a corrosive medium, and analyzed the effects of the nickel layer on coating stress, corrosion current, and crack propagation. The variations in corrosion potential and current were analyzed, and the formation patterns of interfacial corrosion cracks were observed. Pre-cracks were introduced on the ceramic coating surface using a Micro-Nano mechanical testing system, and cyclic impacts were applied to the samples in 5% diluted hydrochloric acid using SiC balls to induce damage evolution. The results indicate that the presence of the nickel interlayer reduced the corrosion current density from 9.197×10^{-6} A/cm^2 to 8.088×10^{-6} A/cm^2 and significantly decreased the stress between the coating and the substrate. The surface cracks gradually extended toward the interface under the coupling effect of corrosion and SiC ball impact. When cracks reached the interface, they provided channels for corrosive media, leading to stress corrosion cracking at the interface. The Ni intermediate layer suppressed the formation of interface cracks and greatly enhanced the impact damage resistance of the CT coating–substrate system in corrosive media.

Keywords: Cr_2O_3-TiO_2; nickel layer; corrosion; cyclic impact; crack evolution

1. Introduction

Ceramic coatings are well known for their excellent high-temperature stability, thermal shock resistance, wear resistance, and optoelectronic properties, particularly excelling in enhancing the wear and corrosion resistance of components [1–4]. Ceramic coatings have been applied to the surfaces of seawater pump blades and deep-sea combustible ice extraction equipment to prevent erosion caused by seawater and sand particles, which are also widely used on the surfaces of fuel pump rotors, mud pump piston rods, and sucker rod couplings in the petroleum industry to enhance wear and corrosion resistance [5–7]. Cr_2O_3 exhibits exceptional hardness, wear resistance, and pressure tolerance. The incorporation of TiO_2 improves its ductility and toughness, resulting in CT coatings with excellent mechanical properties, making them widely used in demanding service environments [8]. However, defects such as porosity and microcracks in the layers are difficult to avoid. Extensive research has been conducted to investigate the microstructure of plasma-sprayed CT ceramic coatings, with particular attention to crack evolution [9]. However, due to the complexity of the internal structure and interface morphology of transition layers and ceramic coatings, studying micro dynamic fracture behavior and how to suppress crack propagation has become a major challenge [10]. Consequently, the inclusion of TiO_2 results in the enhanced toughness and corrosion resistance of the coatings, thus exhibiting exceptional physicochemical characteristics [11].

Microcracks and pores are common issues in plasma-sprayed ceramic coatings [12,13], particularly under cyclic stress conditions, where crack propagation occurs below the fracture strength of the coating [14]. The typical mechanisms of crack propagation involve intrasplat cracking, intersplat decohesion, the interlinking of pores and cracks, mutual splat sliding, and pore compaction, leading to coating failure [15]. Plasma-sprayed coatings are often applied in environments with corrosive media, where corrosion initially occurs at the surfaces of partially unmelted particles, pores, and edges of microcracks [16]. The corrosion mechanism of a ceramic coating–metal substrate system can be attributed to three aspects: (i) penetration of electrolyte and corrosive species through interconnected coating pores towards the substrate, (ii) corrosion of the active area along the coating-substrate interface, and (iii) accumulation of corrosion products at the coating–substrate interface, resulting in a reduction in adhesion strength [17]. In engineering applications, stress corrosion cracking is the primary failure mode of thermal-sprayed ceramic coatings. Coating interfacial stress corrosion cracking (ISCC) refers to the delamination, cracking, and fracture of the coating and substrate at the interface due to stress concentration and the combined effects of corrosion [18]. The mismatch in thermal expansion coefficients between the ceramic coating and the substrate material, as well as the presence of micro-voids at the interface, can trigger stress corrosion cracking, particularly when alternating stress and corrosion act synergistically, promoting crack propagation [19–21].

The impeller blades of seawater pumps and valves of marine equipment are often subjected to erosion from seawater and sand particles. The presence of a high concentration of chloride ions, combined with the cyclic stress caused by sand particle impact, can easily lead to the failure of ceramic coatings. There is a lack of research on the damage evolution law of ceramic coatings under corrosive media and impact loads, and determining how to suppress coating–substrate system failure is an important challenge in engineering. This study conducted particle impact experiments in an environment with a large amount of chloride ions present to analyze the damage evolution mechanism of ceramic coatings. The method of using a nickel layer as a transition layer for the ceramic coating was proposed to investigate the effects of the Ni layer on corrosion potential and current, and to analyze its impact on interface crack propagation. This research holds significant importance for enhancing the service life of ceramic coatings subjected to impact loads in corrosive environments.

2. Experiments

2.1. The Preparation of CT Coatings

In this study, CT coatings with and without a Ni interlayer were prepared on 304 stainless steel using the XM-300SK (Shanghai Xiuma Spraying Machinery, Shanghai, China) supersonic plasma spraying system. The substrate used for this plasma spray was 304 stainless steel with a size specification of $100 \times 100 \times 3$ mm^3. The primary gases and powder carrier gases employed in the system were argon. The transition layer was composed of Ni powders with a particle size of 30–100 μm. TiO_2 accounts for 60% of the total mass of the ceramic powder. The process parameters for CT and the transition layer are shown in Table 1.

Table 1. Plasma process parameters.

Process Parameters	Ni Layer	Ceramic Layer
Current/A	300	500
Voltage/V	62	80
Plasma gas flow (Ar)/L·min^{-1}	60	60
Plasma gas flow (H_2)/L·min^{-1}	10	15
Spray distance (mm)	100	100
Powder feed rate (g/min)	30	30

2.2. Corrosion and Electrochemical Experiments

The CT coating samples after spraying were cut into small pieces of $10 \times 10 \times 3$ mm^3 using a wire cutting machine. After cutting the samples, the cut surfaces were polished and immersed in 5% HCl at room temperature for 1 h to investigate the corrosion behavior of the coating, Ni layer, substrate, and interface. The electrochemical corrosion behaviors were studied by potentiodynamic polarization spectroscopy. Before the experiment, the samples were polished first. The polishing liquid used was a suspension of alumina particles. All the electrochemical experiments were conducted on an electrochemical station (Versa Studio: PARSTAT 3000 A-DX, Berwyn, PA, USA), as shown in Figure 1II. The cell used was a typical three-electrode one fitted with a platinum sheet measuring $1 \times 1 \times 0.05$ cm^3 as the counter electrode (CE), working electrode (specimen) and saturated calomel electrode (SCE) as the reference electrode (RE). The open circuit potential (Ecorr) and corrosion current density (Icorr) of the samples, with and without the Ni layer, were analyzed through electrochemical experiments.

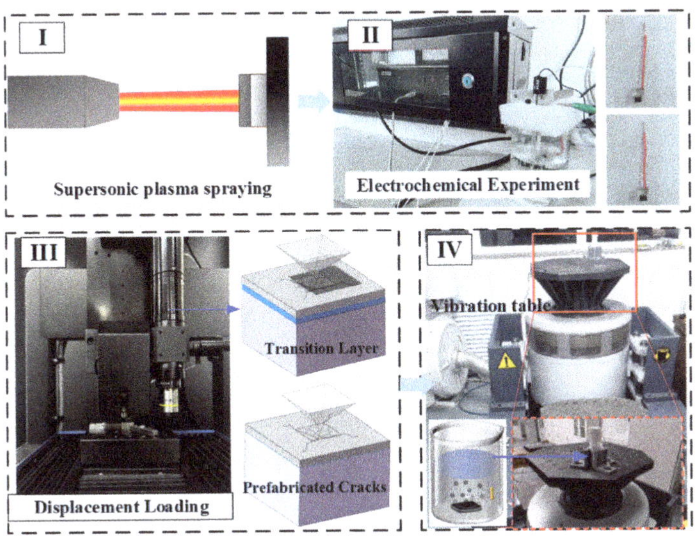

Figure 1. The process of experiments: (**I**) supersonic plasma spraying; (**II**) electrochemical experiments; (**III**) displacement loading experiment; (**IV**) impact experiments in corrosive media.

2.3. Crack Prefabrication

To further study the effect of the Ni layer on interface crack propagation, the Micro-Nano mechanical testing system was used to pre-create indentations and cracks on the ceramic coating surface (Figure 1III). The diamond indenter stayed in situ for 20 s after applying a load of 20 N, forming a diamond indentation on the surface of the coating, and forming cracks in the indentation and at sharp corners.

2.4. Cyclic Impact Experiment in Corrosive Media

To explore the damage evolution of the ceramic coating, a cyclic impact test in a corrosive environment was conducted. The sample was fixed at the bottom of the polytetrafluoroethylene test tube (Figure 1IV), 50 SiC balls with a diameter of 3 mm were added into the test tube, and 5% hydrochloric acid was also added as the corrosion medium. The test tube was installed on the shaking table, and a vibration with an amplitude of 0.5 mm and a frequency of 20 Hz was applied for 2 h. The synergistic effect of cyclic impact stress generated by SiC balls and corrosive media induces the propagation of pre-existing cracks on ceramic coatings.

2.5. Characterization Techniques

After polishing the surface of the coatings, the samples' microstructures were characterized using a scanning electron microscope (SEM, Merlin Compact, Zeiss, Jena, Germany), and the element distribution was examined using energy-dispersive spectroscopy (EDS). The phase composition of the ceramic coatings and substrates was tested using an X-ray diffractometer (XRD, Ultimate IV, RIGAKU, Tokyo, Japan), and the test results were analyzed using MDI Jade 6 software (V.6, ICDD, Philadelphia, PA, USA). Additionally, a simulation was performed to investigate the stress variation pattern in the coatings and analyze the formation mechanism of residual stresses.

3. Results and Discussion

3.1. Microscopic Characteristics of CT Coating–Substrate System

In addition to the abundant presence of chromium (Cr) and nickel (Ni), there are mineral phases of Taenite FeNi and $FeNi_3$ compound phases in the matrix, which can be confirmed by the X-ray diffraction patterns. The appearance of these two phases is attributed to the mutual dissolution, diffusion and reaction between iron and nickel during the heat treatment of the substrate. The internal structure of the coating shown in Figure 2b reveals the presence of Cr_2O_3 and TiO_2, indicating a typical plasma-sprayed ceramic coating. The addition of titanium oxide provides a certain level of toughness to inhibit crack propagation within the coating [22].

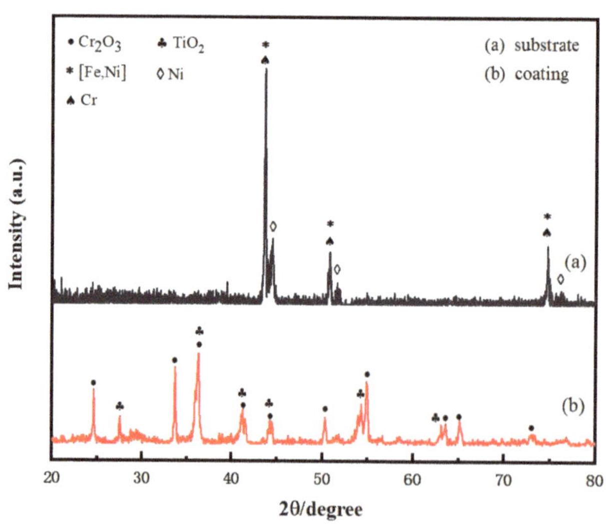

Figure 2. XRD patterns: (**a**) substrate; (**b**) coating.

Powder particles are exposed to high temperatures and kinetic energy through high-pressure gases and heating. When colliding with the substrate surface, the powder particles form a molten or semi-molten splash. These deformed particles intertwine and stack together in a wave-like pattern, forming a layered microstructure, as illustrated in Figure 3a. During the stacking process, the changes in particle flight velocity and temperature triggered the coating structure to show significant irregularities, resulting in gaps or voids between the stacked particles. The elemental distribution map clearly reveals distinct regions and boundaries for Cr and Ti, corresponding to the distribution regions of the particles. The elemental distribution maps clearly show the different regions and boundaries of Cr and Ti, corresponding to the distribution areas of the particles, where the O intensity is significantly higher in the Cr region, indicating higher oxygen content in this region (Figure 2b). The presence of pores and microcracks is commonly observed in ceramic coatings. During plasma spraying, pores and microcracks can be generated from different sources, such as entrapped gases, the incomplete filling in of the rapidly solidifying splats, the shrinking of the splats during rapid solidification, etc. Additionally, as the temperature of deforming particles decreases, they undergo shrinkage. Failure to replenish the liquid phase in a timely manner can also result in the formation of voids within the coating [23–25].

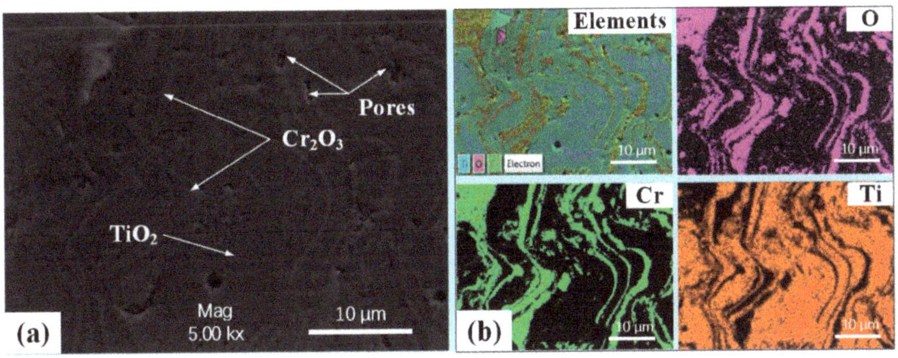

Figure 3. CT coating: (**a**) microstructure of CT coating; (**b**) element distribution.

As shown in Figure 4a, the coating–substrate system was observed to be clearly divided into three parts: the substrate, the transition layer, and the coating. Interface 1 and Interface 2 were formed between the nickel layer, ceramic coating, and substrate, achieving a strong bond between the materials. The elemental distribution of the coating–substrate system was investigated using energy-dispersive X-ray spectroscopy (EDS, Zeiss Merlin Compact, Zeiss, Jena, Germany), revealing distinct separation into three different regions. The transition layer served to bind the coating and the substrate, and was mainly composed of nickel. However, the EDS spectrum reveals the presence of Al and O in the transition layer, which is attributed to the retention of alumina particles during metallographic polishing. The ceramic coating primarily contained elements such as Cr, Ti, and O. The substrate was mainly composed of elements such as Fe and Ni. Figure 4b shows numerous black particles embedded in the transition layer. EDS analysis identifies these black particles as corresponding to aluminum elements, confirming that they are alumina abrasive particles embedded in the nickel layer.

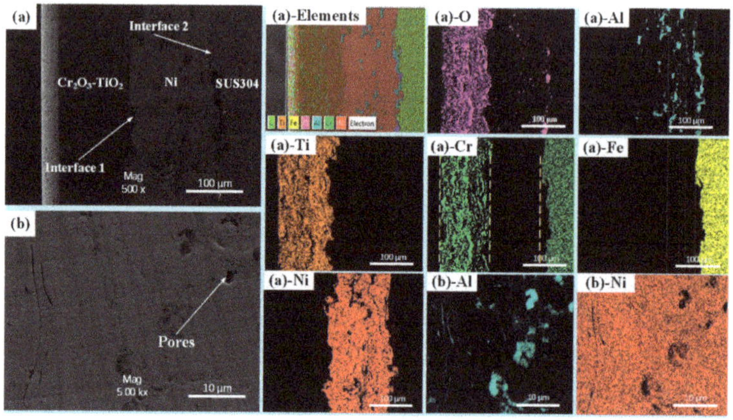

Figure 4. Coating–substrate system and element distribution: (**a**) ceramic coating with transition layer; element distribution (**b**) transition layer; element distribution.

3.2. Interfacial Crack Formation

To investigate the mechanism by which a nickel layer influences the interfacial cracking of CT coatings, two types of coating–substrate systems were prepared using plasma spraying technology. In the first type of sample, the CT coating was directly sprayed onto the 304 stainless steel. In the second type, a nickel layer was deposited between the substrate and the CT coating. The coating exhibited numerous pores, with a particularly high density of pores and microcracks at the interface, as shown in Figure 5a. This phenomenon occurred because, upon the impact of a droplet on a solid surface, a thin air film forms between the center of the droplet and the substrate. This air film ultimately evolves into a spherical bubble to minimize surface energy, resulting in the formation of pores at the interface [25]. Additionally, during the cooling process, thermal stresses arise at the interface due to the different thermal expansion coefficients. These stresses generate interface cracks originating from defects such as pores [26]. The samples were placed in a 5% HCl solution for a corrosion period of 2 h at room temperature. As shown in Figure 5b, corrosion occurred at the interface between the substrate and the coating. The corrosion traces on the substrate exhibited a striped pattern, while pronounced continuous cracks were observed at the interface. This confirms that the interface is the most vulnerable part of the thermal-sprayed coating in a corrosive environment, and corrosion-induced cracking at the interface is an important reason for the failure of the thermal-sprayed coating. Elemental distribution analysis reveals the presence of elements such as O, Cr, and Ti in the coating, while showing that the substrate predominantly contains Fe. A clear demarcation is observed between the coating and the substrate.

Corrosion experiments on samples with a nickel layer revealed that the nickel layer and ceramic coating were well bonded, with no cracks observed at Interface 1. However, significant cracks were found at Interface 2, between the nickel layer and the substrate, as shown in Figure 6a. Figure 6b,c show numerous pores in both the ceramic coating and the nickel layer. Interface 1 is clearly identified and surrounded by many defects. Obvious corrosion cracks were formed at Interface 2, and obvious β-phase corrosion occurred on the substrate. However, there were fewer corrosion marks on the ceramic coating and nickel layer, and slight intergranular corrosion was observed on the nickel layer (Figure 6d–f). As shown in Figure 6j, elemental distribution indicates that the transition layer is predominantly composed of nickel, while aluminum is introduced from alumina abrasive particles in the polishing solution. The ceramic coating depicted in Figure 6i contains elements such as Ti, Cr, and O, indicating that it is a corrosion-resistant ceramic.

The intermediate layer is primarily composed of nickel, while the aluminum element is introduced from the alumina abrasive particles in the polishing solution (Figure 6j). The enhanced corrosion resistance at Interface 1 inhibits the growth of interfacial cracks, while the cracks at Interface 2 are formed due to the corrosion of the substrate near the interface.

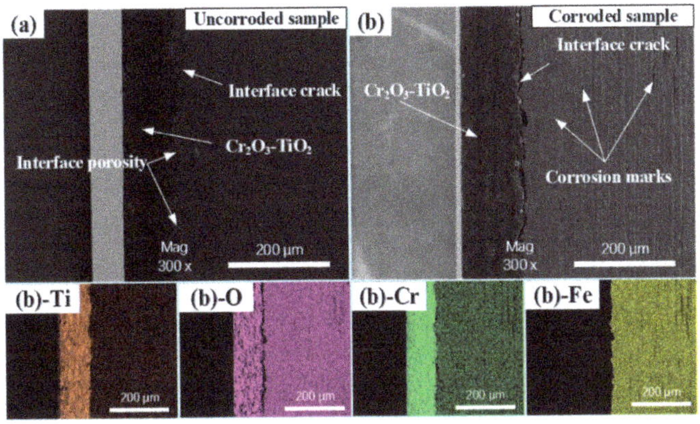

Figure 5. Corrosion cracking of the coating interface: (**a**) uncorroded sample; (**b**) corroded sample and its element distribution.

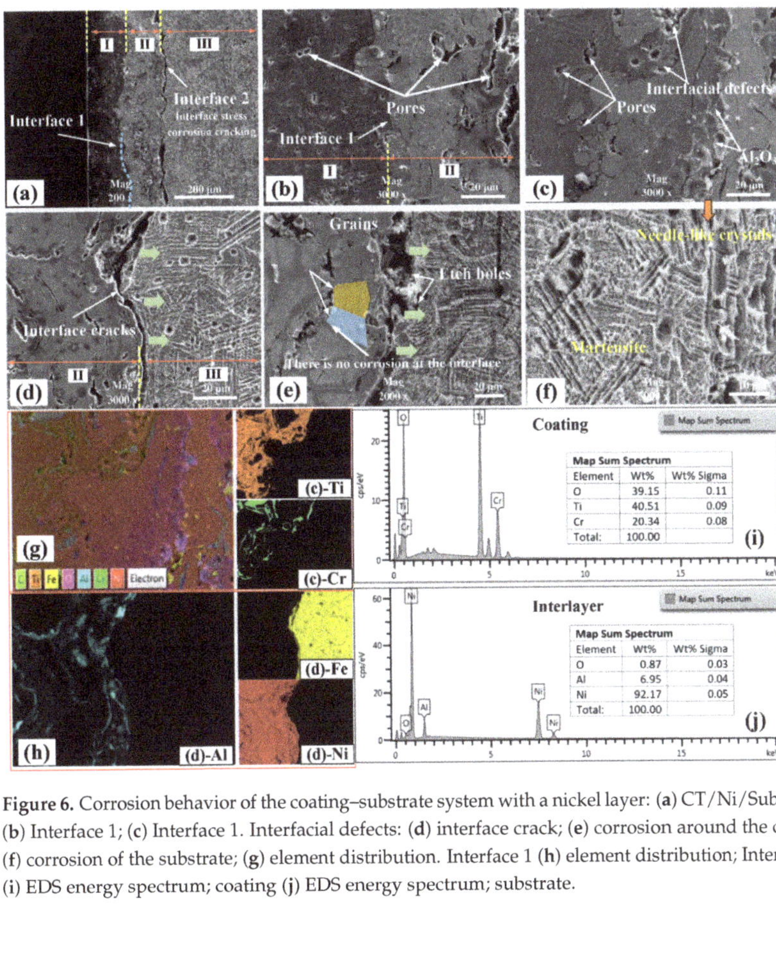

Figure 6. Corrosion behavior of the coating–substrate system with a nickel layer: (**a**) CT/Ni/Substrate; (**b**) Interface 1; (**c**) Interface 1. Interfacial defects: (**d**) interface crack; (**e**) corrosion around the cracks; (**f**) corrosion of the substrate; (**g**) element distribution. Interface 1 (**h**) element distribution; Interface 2 (**i**) EDS energy spectrum; coating (**j**) EDS energy spectrum; substrate.

3.3. Electrochemical Characteristics

To evaluate the corrosion resistance of CT/Ni/Substrate and CT/Substrate, Tafel curves of the two samples were obtained using an electrochemical workstation. As depicted in Figure 7, the passivation region of the sample with the added nickel layer is more pronounced compared to that of the CT/Substrate sample, indicating the easier formation of a passive film and enhanced corrosion resistance. Furthermore, the corrosion potential of the CT/Ni/Substrate sample increased from −0.657 V to −0.766 V, and the corrosion current density decreased from 9.197×10^{-6} A/cm^2 to 8.088×10^{-6} A/cm^2. The negative shift in potential is merely a thermodynamic tendency. The decrease in current density demonstrates the enhanced corrosion resistance of the CT/Ni/Substrate sample. Additionally, the passivation region of the CT/Ni/Substrate samples was significantly enlarged, indicating that the incorporation of the Ni layer facilitates the formation of a passivation film on the surface, thereby enhancing the passivation phenomenon.

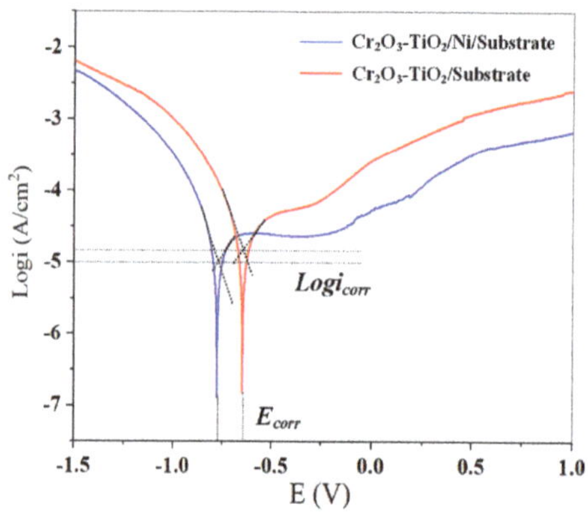

Figure 7. Polarization curve of CT/Ni/Substrate and CT/Substrate.

3.4. Damage Evolution of the Coating–Substrate System

Vibrations with a frequency of 20 Hz and an amplitude of 0.5 mm were generated using a vibration table, causing SiC particles to repeatedly impact the ceramic coating surface in a 5% hydrochloric acid solution. Figure 8a illustrates two types of failure modes occurring on the coating surface. Cracking is a common and almost unavoidable failure mode for ceramic coatings, especially under the combined effects of corrosive media and cyclic stress, leading to crack propagation on the surface. As cracks propagate, corrosion continues within the crack gaps, which is associated with stress corrosion cracking. The susceptibility to stress corrosion cracking is significantly higher than it is under non-corrosive or non-stress conditions [27–29]. Figure 8d shows corrosion and minor spalling on both sides of the cracks. The other form of damage is the localized spalling of the coating, manifested as the layer-by-layer delamination of the ceramic coating, initiated by defects such as pores, eventually forming deep pits on the coating surface, as depicted in Figure 8b,c.

To further investigate the damage evolution patterns of the film–substrate system, indents and pre-cracks were prepared on two samples: CT/Ni/Substrate and CT/Substrate. The displacement–load curves of the two samples are similar. However, under the same applied load of 20 N, the displacement of the CT/Ni/Substrate sample is generally larger than that of the CT/Substrate sample (Figure 9). This is because when the indentation depth exceeds 1/10 of the coating thickness, the resulting displacement–load curve represents the film–substrate system rather than the ceramic coating alone. The nickel layer exhibits good plasticity, leading to more significant deformation under the same load conditions. Consequently, the indentation depth is slightly deeper.

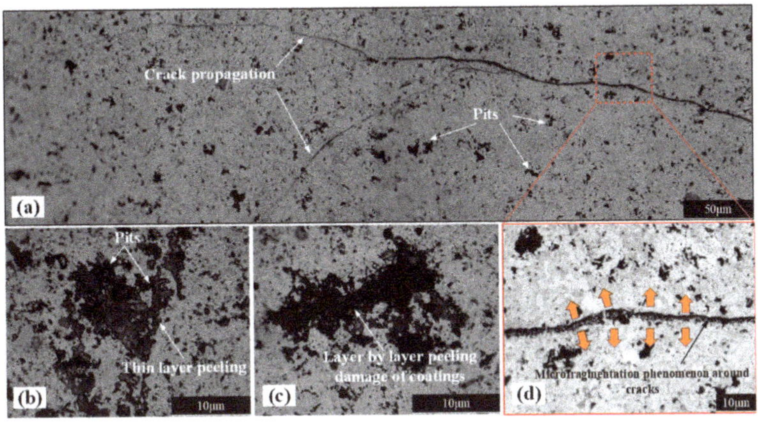

Figure 8. Surface damage forms of CT coatings.

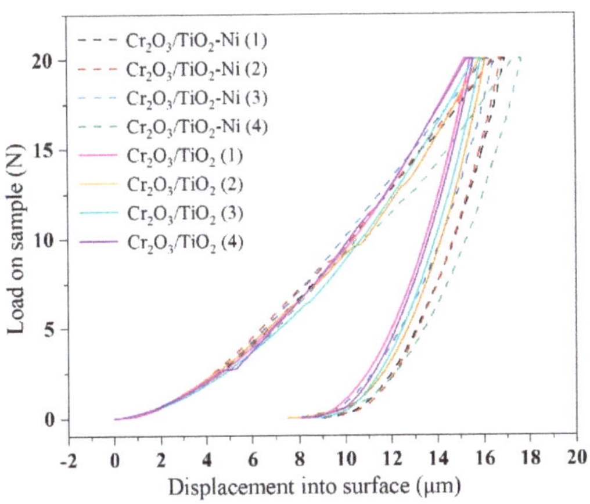

Figure 9. Displacement–load relationship curve.

The indenter of the Micro-Nano mechanical testing system induces two types of damage: indentations and cracks. Figure 10aI shows the indentation of the CT/Ni/Substrate sample, where cracks are generated at the sharp corners of the indentation, and there is slight spalling of the coating at the edges. As the corrosion medium impact experiment progresses, the spalled area of the coating increases, and the cracks within the indentation interweave in a spiderweb-like pattern (Figure 10b2). After 2 h of the corrosion impact experiment, localized delamination of the coating within the indentation occurs, as shown in Figure 10aIII. Figure 10b1 displays the indentation of the CT/Substrate sample. Without the nickel layer, the ceramic coating is more prone to cracking and large-scale delamination. This phenomenon is likely related to the internal stress within the coating and the bonding strength between the film and the substrate. As the impact experiment continues, delamination begins within the coating, and extensive delamination occurs around the warped coating (Figure 10b2). Without a nickel layer as a barrier, the interaction of the corrosion medium and abrasive particles results in the almost complete peeling of the coating within the indentation, deepening the corrosion and continuously increasing the peeling area.

These observations indicate that the presence of the nickel layer effectively suppresses the deterioration of the coating damage.

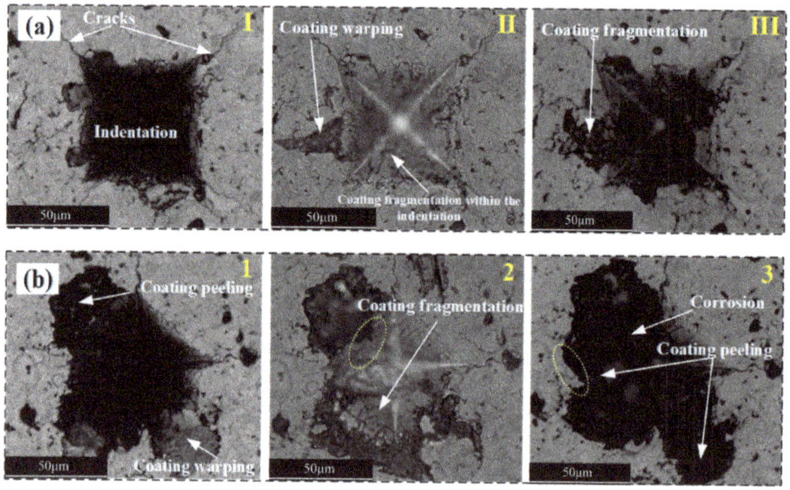

Figure 10. Prefabricated damage and its evolution patterns in a corrosive impact environment: (**a**) CT/Ni/Substrate; (**b**) CT/Substrate.

The phenomenon of cracking induced by the combined effect of lower stress (below the fatigue strength) and a mild corrosive environment is referred to as stress corrosion cracking. If only one factor, either stress or the corrosive medium, is present, failure does not occur. However, when both factors act together, cracking can occur rapidly [30]. Stress corrosion leads to a reduction in fracture toughness, which is attributed to the presence of corrosive media. These media attack interlayer bonding and weaken the bond energy of closely connected chemical bonds, thereby causing damage to the crack tip model and initiating rapid crack propagation. Stress plays a crucial role in crack propagation [31]. As shown in Figure 11a, the stress base along the X-axis and Y-axis exhibits tensile stress, while compressive stress is generated within the coating. In the Z-axis direction, tensile stress is present in the areas with the coating. After the introduction of the Ni layer, the stress within both the substrate and the coating decreases; however, stress concentration is observed in the X-axis direction of the Ni layer, and compressive stress is present in the Z-axis direction. In Figure 11b, the CT/Ni/Substrate sample coating exhibits compressive stress in the X-axis direction, which is lower than that of the CT/Substrate sample. The maximum tensile stress occurs at Interface 2, and it is significantly lower than that of the CT/Substrate sample. The presence of the Ni layer transforms the stress in the Z-axis direction from tensile to compressive (Figure 11c). The formation of interfacial cracks aligns with the stress concentration at Interface 2. The development of compressive stress within the CT/Ni/Substrate sample coating and the overall reduction in stress contribute to fewer cracks within the coating, as illustrated in Figure 11b.

The combined effects of corrosive media corrosion and the cyclic impact of SiC balls induce the formation of cracks in ceramic coatings. The propagation of these cracks occurs in two forms: one is parallel to the coating surface, leading to the layer-by-layer spalling of the ceramic coating, and the another form is a through-thickness crack, characterized by its extension towards the coating–substrate interface, as illustrated in Figure 12a. The corrosive medium infiltrates the membrane–substrate interface through capillary action via cracks generated during fatigue cycles or pores in the surface coating of the sample, leading to the formation of corrosion pits. Subsequently, these pits initiate and propagate cracks within the substrate [32,33]. The micromechanical system generates surface cracks, which progressively extend towards the interface under the repeated impacts of the corrosive media and SiC balls. When the cracks reach the interface, they provide a pathway for the corrosive media, resulting in stress corrosion cracking at the interface [34], as shown in Figure 12b. Figure 12d reveals that the substrate is preferentially corroded by the media, and the occurrence of interface cracks depends on the substrate's corrosion resistance. After the addition of a nickel layer (Figure 12c), when the cracks extend to the interface, the corrosive media contact the nickel layer, inhibiting corrosion and making it difficult for interface cracks to form (Figure 12c,e,f). Furthermore, the presence of the nickel layer effectively reduces the stress at the interface, significantly enhancing the suppression of stress corrosion cracking at the interface. Consequently, the constructed CT/Ni/304 stainless steel coating–substrate system exhibits excellent performance in suppressing damage evolution under corrosive media and repeated impact conditions.

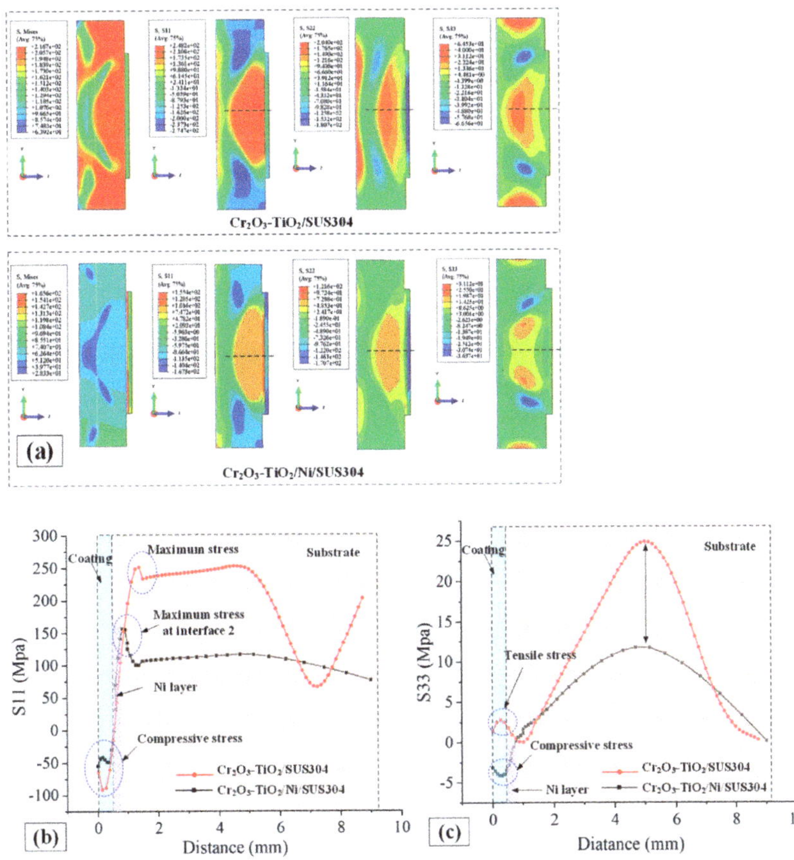

Figure 11. Stress distribution characteristics. (**a**) Stress cloud maps of two samples: (**b**) S11 and (**c**) S33.

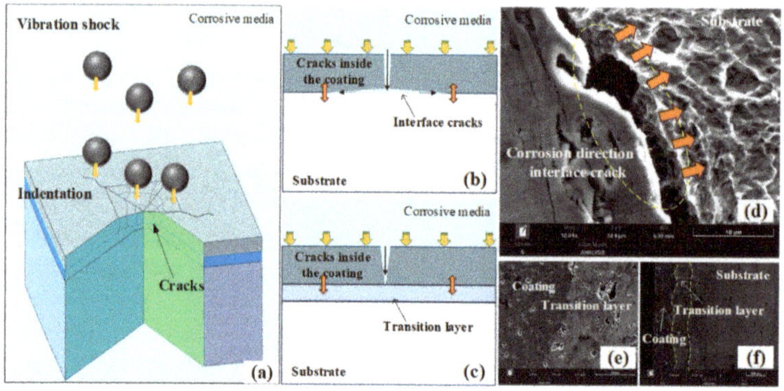

Figure 12. Damage evolution in corrosive and impact environments: (**a**) damage generation; (**b**) CT/Substrate; (**c**) CT/Ni/Substrate; (**d**) interface cracks; (**e**) interface between nickel layer and ceramic coating; (**f**) CT/Ni/Substrate.

4. Conclusions

To enhance the suitability of plasma-sprayed ceramic coatings under corrosive impact conditions, this study deposited a nickel layer between the ceramic coating and 304 stainless steel. The findings demonstrate that the nickel layer reduces interfacial and coating stresses, suppresses the propagation of interfacial cracks, and effectively improves the damage resistance of the ceramic coating.

(a) The formation of interfacial cracks primarily depends on the corrosion resistance of the materials on either side of the interface. Therefore, the addition of a nickel layer effectively inhibits the formation of cracks between the ceramic coating and the nickel layer. However, it cannot prevent the formation of cracks between the nickel layer and the substrate, which is due to the substrate's poor corrosion resistance.
(b) The nickel layer reduces the stress on the coating and substrate, but stress concentration occurs near the interface between the nickel layer and substrate. The stress in the Z-axis direction changes from tensile stress to compressive stress.
(c) The presence of the nickel layer enhances the corrosion resistance of the coating–substrate system and strengthens the passicvation phenomenon. The corrosion potential of the CT/Ni/Substrate sample increased from −0.657 V to −0.766 V, and the corrosion current density decreased from 9.197×10^{-6} A/cm^2 to 8.088×10^{-6} A/cm^2.
(d) The combined effects of corrosion and impact result in two types of damage on the surface of ceramic coatings: through-thickness cracks and corrosion pits. Through-thickness cracks are the primary cause of interfacial cracks, while corrosion pits arise from the progressive delamination of the ceramic coating layers.

Author Contributions: H.Y.: methodology, experiment, data curation, and writing—original draft. Y.Z.: experiment, writing—review and editing. X.Q.: simulation. Y.J.: writing—review and editing. X.Z.: experiment. K.T.: investigation and simulation. X.W.: methodology. Z.Z.: conceptualization and supervision. All authors have read and agreed to the published version of the manuscript.

Funding: This project was supported by National Natural Science Foundation of China (Grant No. 52305510), National Defense Science and Technology Key Laboratory (6142005KJW2022011) and Natural Science Foundation of Shandong Province (ZR2023QE044).

Data Availability Statement: All data and material are presented in the main manuscript and additional supporting files are available from corresponding author on reasonable request.

Conflicts of Interest: The authors declare that they have no conflicts of interest.

References

1. Han, C.; Wang, Z.; Ma, L.; Huang, G.; Ma, Y. Cold spray for ceramic metallisation: A review. *Adv. Appl. Ceram.* **2021**, *120*, 358–380. [CrossRef]
2. Missner, M.; Stryhalski, J.; Tomiyama, M.; Soares, P.; Recco, A.; Fontana, L. Rutile TiO_2 thin films growth on glass substrates with generation of high entropy interface. *J. Mater. Res. Technol.* **2023**, *24*, 963–970. [CrossRef]
3. Talavari, F.; Shakeri, A.; Mighani, H. Synthesis of Cr_2O_3/TiO_2 nanocomposite and its application as the blocking layer in solar cells. *J. Environ. Anal. Chem.* **2018**, *5*, 2380–2391. [CrossRef]
4. Dey, A.; Banerjee, K.; Mukhopadhyay, A.K. Microplasma sprayed hydroxyapatite coating: Emerging technology for biomedical application. *Mater. Technol.* **2014**, *29*, 15–20. [CrossRef]
5. Zhang, H.; Wang, Y.; Chen, X.; Zhang, Z.; Zeng, X.; Cheng, X. Corrosion behaviors of Al_2O_3–20TiO_2 and Cr_2O_3–3TiO_2–5SiO_2 coatings in both artificial seawater and high-pressure hydrogen sulfide seawater. *Ceram. Int.* **2024**, *50*, 34346–34356. [CrossRef]
6. Lakshmi, K.P.J.; Irappa, S.; Raghavendra, C.R.; Basavarajappa, S. High temperature erosive behaviour of plasma sprayed NiCrAlY/B4C/Cenospherecoating on MDN 321 turbine steel. *Int. J. Surf. Eng. Coat.* **2023**, *101*, 49–56.
7. Zavareh, M.A.; Sarhan, A.A.D.M.; Karimzadeh, R.; Singh, R.S.A. Analysis of corrosion protection behavior of Al_2O_3-TiO_2 oxide ceramic coating on carbon steel pipes for petroleum industry. *Ceram. Int.* **2018**, *44*, 5967–5975. [CrossRef]
8. Vernhes, L.; Bekins, C.; Lourdel, N.; Poirier, D.; Lima, R.S.; Li, D.; Klemberg-Sapieha, J.E. Nanostructured and conventional Cr_2O_3, TiO_2, and TiO_2-Cr_2O_3 thermal-sprayed coatings for metal-seated ball valve applications in hydrometallurgy. *J. Therm. Spray Technol.* **2016**, *25*, 1068–1078. [CrossRef]
9. Rasooli, A.; Safavi, M.S.; Hokmabad, M.K. Cr_2O_3 nanoparticles: A promising candidate to improve the mechanical properties and corrosion resistance of Ni-Co alloy coatings. *Ceram. Int.* **2018**, *44*, 6466–6473. [CrossRef]
10. He, X.; Song, P.; Huang, T.; Wan, F.; Kong, D.; Zhai, R.; Hua, C.; Dai, J. Fracture process simulation and crack resistance behavior analysis of transition-layer ceramic coating based on real image reconstruction model. *Surf. Interfaces* **2024**, *46*, 104003. [CrossRef]
11. Babu, P.S.; Sen, D.; Jyothirmayi, A.; Krishna, L.R.; Rao, D.S. Influence of microstructure on the wear and corrosion behavior of detonation sprayed Cr_2O_3-Al_2O_3 and plasma sprayed Cr_2O_3 coatings. *Ceram. Int.* **2018**, *44*, 2351–2357. [CrossRef]
12. Wang, S.; Liu, X.; Zhang, J.; Xu, P.; Wei, M.; Liu, G.; Zhan, X.; Coyle, T.W.; Mostaghimi, J. Tailoring the wetting behaviors and surface structures of yttrium oxide coatings deposited via different plasma spray processes. *J. Mater. Res. Technol.* **2024**, *29*, 1924–1936. [CrossRef]
13. Ahmadian, S.; Jordan, E.H. Explanation of the effect of rapid cycling on oxidation, rumpling, microcracking and lifetime of air plasma sprayed thermal barrier coatings. *Surf. Coat. Technol.* **2014**, *244*, 109–116. [CrossRef]
14. Kovářík, O.; Haušild, P.; Medřický, J.; Tomek, L.; Siegl, J.; Mušálek, R.; Curry, N.; Björklund, S. Fatigue Crack Growth in Bodies with Thermally Sprayed Coating. *J. Therm. Spray Technol.* **2016**, *25*, 311–320. [CrossRef]
15. Mušálek, R.; Kovářík, O.; Matějíček, J. In-situ observation of crack propagation in thermally sprayed coatings. *Surf. Coat. Technol.* **2010**, *205*, 1807–1811. [CrossRef]
16. Huang, B.; Wang, J.; Tang, Z.-H.; Li, W.-D.; Zhu, W.-J.; Gu, R.-B. Fluoride-mediated corrosion mechanism of atmospheric-plasma-sprayed yttrium–aluminium garnet ceramic coatings. *J. Eur. Ceram. Soc.* **2022**, *42*, 6146–6158. [CrossRef]
17. Zamani, P.; Valefi, Z.; Jafarzadeh, K. Comprehensive study on corrosion protection properties of Al_2O_3, Cr_2O_3 and Al_2O_3–Cr_2O_3 ceramic coatings deposited by plasma spraying on carbon steel. *Ceram. Int.* **2022**, *48*, 1574–1588. [CrossRef]
18. Chen, W.; Hu, T.; Wang, C.; Xiao, H.; Meng, X. The effect of microstructure on corrosion behavior of a novel AlCrTiSiN ceramic coating. *Ceram. Int.* **2020**, *46*, 12584–12592. [CrossRef]
19. Zhang, Z.; Zhong, X.; Li, L.; Hu, J.; Peng, Z. Unmasking the delamination mechanisms of a defective coating under the co-existence of alternating stress and corrosion. *Prog. Org. Coat.* **2023**, *180*, 107560. [CrossRef]
20. Ozkan, D. Structural characteristics and wear, oxidation, hot corrosion behaviors of HVOF sprayed Cr_3C_2-NiCr hardmetal coatings. *Surf. Coat. Technol.* **2023**, *457*, 129319. [CrossRef]
21. Wang, Z.; Zhang, J.; Liu, X.; Liu, J.; Liu, G. Failure mechanism of plasma sprayed Tb-YSZ coating under NaCl high-low temperature cyclic corrosion. *Surf. Coat. Technol.* **2023**, *456*, 129244. [CrossRef]
22. Hanifi, D.; Hakimi, N.; Ali, R.; Huang, T.; Huma, T.; Bakhshyar, D.; Afzali, N.R.; Shafi, M.; Babeker, H.; Lu, J.; et al. The fracture toughness and tribological performance of NiAl/Al_2O_3-40 Wt.% TiO_2 coating generated by air plasma spraying. *Ceram. Int.* **2024**, *50*, 1533–1546. [CrossRef]
23. Thirumalaikumarasamy, D.; Shanmugam, K.; Balasubramanian, V. Effect of atmospheric plasma spraying parameters on porosity level of alumina coatings. *Surf. Eng.* **2012**, *28*, 759–766. [CrossRef]
24. Ctibor, P.; Lechnerová, R.; Beneš, V. Quantitative analysis of pores of two types in a plasma-sprayed coating. *Mater. Charact.* **2006**, *56*, 297–304. [CrossRef]
25. Shukla, R.K.; Kumar, A.; Kumar, R.; Singh, D.; Kumar, A. Numerical study of pore formation in thermal spray coating process by investigating dynamics of air entrapment. *Surf. Coat. Technol.* **2019**, *378*, 124972. [CrossRef]

26. Ferguen, N.; Leclerc, W.; Lamini, E.-S. Numerical investigation of thermal stresses induced interface delamination in plasma-sprayed thermal barrier coatings. *Surf. Coat. Technol.* **2023**, *461*, 129449. [CrossRef]
27. Terán, G.; Capula-Colindres, S.; Velázquez, J.; Zuñiga-Hinojosa, M.; Contreras, A. Modeling 3D finite element analysis of a semi-elliptical crack on stress corrosion cracking of API X52 pipeline. *Forces Mech.* **2024**, *16*, 100279. [CrossRef]
28. Han, C.; Yu, M.; Jia, X.; Zhao, Z.; Zhang, C.; Xiao, J.; Li, S.; Liu, J. Effect of inhomogeneous microstructures on stress corrosion cracking sensitivity of 2050-T84 Al-Li alloy thick plate. *Corros. Sci.* **2024**, *235*, 112213. [CrossRef]
29. Wei, Z.-Y.; Cai, H.-N. Stress states and crack behavior in plasma sprayed TBCs based on a novel lamellar structure model with real interface morphology. *Ceram. Int.* **2019**, *45*, 16948–16962. [CrossRef]
30. Loto, C.A. Stress corrosion cracking: Characteristics, mechanisms and experimental study. *Int. J. Adv. Manuf. Technol.* **2017**, *93*, 3567–3582. [CrossRef]
31. Zhang, W.; Sun, J.; Ding, D.; Hou, D. Nanoscale insights on the stress corrosion mechanism of calcium-silicate-hydrate. *J. Build. Eng.* **2023**, *80*, 107991. [CrossRef]
32. Tokaji, K.; Ogawa, T.; Hwang, J.U.; Kobayashi, Y.; Harada, Y. Corrosion Fatigue Behavior of a Steel with Sprayed Coatings. *J. Therm. Spray Technol.* **1996**, *5*, 269–276. [CrossRef]
33. Morelli, S.; Bursich, S.; Testa, V.; Bolelli, G.; Miccichè, A.; Lusvarghi, L. CMAS corrosion and thermal cycling fatigue resistance of alternative thermal barrier coating materials and architectures: A comparative evaluation. *Surf. Coat. Technol.* **2022**, *439*, 128433. [CrossRef]
34. Cai, H.; Shan, X.; Lu, J.; Luo, L.; Cai, Z.; Wang, W.; Zhang, X.; Zhao, X. The crack behavior and delamination mechanisms of air plasma sprayed thermal barrier coatings under ultrasonic plasma jet at 1600 °C. *J. Eur. Ceram. Soc.* **2023**, *43*, 4136–4145. [CrossRef]

Disclaimer/Publisher's Note: The statements, opinions and data contained in all publications are solely those of the individual author(s) and contributor(s) and not of MDPI and/or the editor(s). MDPI and/or the editor(s) disclaim responsibility for any injury to people or property resulting from any ideas, methods, instructions or products referred to in the content.

coatings

MDPI

Article

Tuning Nanocrystalline Heterostructures for Enhanced Corrosion Resistance: A Study on Electrodeposited Ni Coatings

Wenyi Huo [1,2,*], Zeling Zhang [3], Xuhong Huang [1], Yueheng Wang [1], Shiqi Wang [4], Xiaoheng Lu [1], Shuangxiao Li [1], Senlei Zhu [1], Feng Fang [3] and Jianqing Jiang [1,3]

1 College of Mechanical and Electrical Engineering, Nanjing Forestry University, Nanjing 210037, China
2 NOMATEN Centre of Excellence, National Centre for Nuclear Research, 05-400 Otwock, Poland
3 Jiangsu Key Laboratory of Advanced Metallic Materials, Southeast University, Nanjing 211189, China
4 Department of Chemistry, University of Helsinki, P.O. Box 55, 00014 Helsinki, Finland
* Correspondence: wyhuo@njfu.edu.cn

Abstract: Tailoring the microstructural heterogeneity of metallic coatings is a promising strategy for enhancing their corrosion resistance; however, its systematic optimization remains underexplored. Here in, we present a one-step, scalable electrodeposition strategy to fabricate Ni coatings with tunable nanocrystalline heterostructures on Cu substrates by varying the current density from 1 mA/cm^2 to 50 mA/cm^2. The coating with a current density of 10 mA/cm^2, featuring a heterogeneous nanograin structure of coexisting small and large grains, exhibited optimal corrosion resistance in 3.5 wt.% NaCl solution, with a low self-corrosion current density of 4.48 $\mu A/cm^2$. Electrochemical impedance spectroscopy (EIS) and molecular dynamics (MD) simulations revealed that the heterostructure dispersed Cl^- adsorption sites and promoted passivation. High-resolution transmission electron microscopy (HRTEM) revealed that as the current density increased from 10 mA/cm^2 to 50 mA/cm^2, the corrosion product transitioned from a crystalline NiOOH structure to an amorphous structure, which correlated with a reduced corrosion resistance. The heterogeneous microstructure enhances durability, offering a cost-effective and alloy-free alternative for offshore applications. These findings provide a theoretical and experimental basis for designing advanced corrosion-resistant coatings.

Keywords: Ni coatings; electrodeposition; heterogeneous structure; compositional plainification; corrosion resistance; Cl ion adsorption

1. Introduction

Ni-based coatings have been widely utilized in engineering applications, particularly in offshore environments, owing to their corrosion resistance and mechanical reliability [1,2]. However, the increasing demands of modern engineering, such as extended service life, resistance to corrosion, and cost-effectiveness, present challenges that conventional Ni-based coatings may struggle to address fully. To overcome these shortcomings, an effective strategy is compositional plainification [3], a design philosophy that improves material performance by microstructural optimization alone [4], avoiding the complexity and cost of alloying elements [5]. Unlike alloyed coatings, such as high-entropy alloys, which depend on compositional diversity to enhance corrosion resistance at the cost of processing complexity and resource demands, compositional plainification uses controlled grain size refinement and structural heterogeneity to achieve superior properties [6–8]. Recent advances have highlighted nanostructured coatings with heterogeneous microstructures [9], which are promising alternatives. These heterostructured materials exhibit synergistic properties, including enhanced mechanical strength [10], oxidation resistance [11], and corrosion resistance [12]. These properties make them suitable candidates for next-generation coatings in harsh environments.

Electrodeposition is an advantageous technique for synthesizing nanostructured coatings. It provides a scalable, single-step process for customizing microstructural features, such as grain size and phase distribution [13,14]. Electrodeposition has been successfully applied to fabricate Ni-based coatings, including pure Ni and its binary to quinary alloys [15–18]. However, its potential for producing nanocrystalline heterostructures remains unexplored. Previous studies on electrodeposition provide a foundation for this work. Studies on electrodeposited Ni-based alloys have demonstrated that deposition parameters, such as current density, electrolyte composition, and temperature, influence grain refinement and coating properties [19–22]. Among these, our earlier investigation showed that tuning the current density is a straightforward and customizable approach to introducing heterostructures into electrodeposited coatings [16]. However, a systematic understanding of how these parameters control the formation of heterostructured coatings and their corrosion mechanisms remains limited. This gap implies the need for a comprehensive approach that integrates experimental synthesis with thorough characterization and corresponding simulations.

The approach in the current work yields Ni coatings with tunable microstructures designed to meet the rigorous demands of offshore engineering. To clarify the relationship between nanoscale architecture and corrosion resistance, the samples were characterized using a suite of experimental techniques, including X-ray diffraction (XRD), scanning electron microscopy (SEM), and electrochemical tests. The adsorption mechanisms governing the corrosion behavior were simulated using molecular dynamics (MD). The corrosion products were characterized using high-resolution transmission electron microscopy (HRTEM) to reveal the transitions in the corrosion mechanism. By linking the interplay between structural heterogeneity and corrosion resistance, this work provides actionable insights into the design of durable and cost-effective coatings. These findings offer a pathway for the development of durable coatings tailored for offshore energy and chemical engineering.

2. Materials and Methods

2.1. Electrodeposition Preparation

Pure Cu (≥99.9%) was selected as the substrate material. The substrates were machined into small pieces (15 mm × 15 mm × 1 mm), polished with SiC papers and diamond slurry, and etched in 10 vol. % sulfuric acid for 30 s, rinsed with ultrapure water, and dried with nitrogen gas. During this process, the surface was degreased by ultrasonic cleaning. Nanocrystalline Ni coatings were electrodeposited on pure copper using the direct current method at room temperature [16]. Each liter of the electrolytic bath solution comprised 76.1 mM $NiSO_4$, 466.4 mM $C_6H_8O_7$, 163.2 mM $C_6H_5Na_3O_7$, 1368.9 mM NaCl, 404.3 mM H_3BO_3, 7.3 mM $C_7H_4NNaO_3S$, and 0.3 mM $C_{12}H_{25}SO_4Na$ (pH = 2.5). The electrolytic bath was prepared using ultrapure deionized water. A square platinum sheet electrode (1 cm × 1 cm) was used as the anode. The electrodeposition time was 30 min for each sample. During electrodeposition, different current densities were applied: 1 mA/cm^2, 5 mA/cm^2, 10 mA/cm^2, 20 mA/cm^2 and 50 mA/cm^2. Based on the differences between the pre-and post-deposition weights, the estimated thicknesses for current densities of 1 mA/cm^2, 5 mA/cm^2, 10 mA/cm^2, 20 mA/cm^2, and 50 mA/cm^2 were 0.3 μm, 1.9 μm, 3.7 μm, 6.2 μm, and 13.4 μm, respectively.

2.2. Material Characterization

The phase structure of the electrodeposited Ni coatings was identified using XRD (Ultima IV, Rigaku Corporation, Tokyo, Japan). Cu Kα radiation was used for the measurements. The operating voltage and current were 40 kV and 30 mA, respectively. The surface morphology was observed using field emission SEM (FE-SEM, Nova NanoSEM 450, FEI, Hillsboro, OR, USA) at an operating voltage of 15 kV. The electrochemical properties were evaluated using an electrochemical workstation (CHI-660E, Chenhua, Shanghai, China) with a standard three-electrode configuration. During the measurements, the test medium was a 3.5 wt.% NaCl solution at room temperature. The three-electrode system consisted of the prepared electrodeposited Ni coating as the working electrode, a saturated calomel electrode (SCE) as the reference electrode, and a platinum mesh electrode as the counter electrode. The exposed area was 1.0 cm^2. Before the electrochemical measurements, the open-circuit potential (OCP) of each sample was monitored for 10 min. After the OCP measurements, the potential stabilized for all samples, confirming a steady state for subsequent electrochemical impedance spectroscopy (EIS) and polarization tests. During the EIS test, the frequency was set within a range from 0.1 Hz to 10 MHz, the sinusoidal perturbation was 5 mV, and the spectroscopy data were obtained at 0 V. After the EIS tests, the data were fitted using the ZSimpWin 3.30d software. Potentiodynamic polarization tests were performed at a scanning rate of 5.0 mV/s to evaluate the uniform corrosion of the coating. The results were further analyzed using the Tafel extrapolation method. To understand the corrosion products, freestanding TEM samples were mechanically peeled off from the samples after polarization testing. HRTEM tests were carried out using scanning TEM (STEM, Talos F200X, Thermo Fisher Scientific, Waltham, MA, USA) at an operating voltage of 200 kV. Energy dispersive spectroscopy (EDS, X-Max, Oxford Instruments, High Wycombe, UK) was used to analyze element distribution.

2.3. Adsorption Simulation

MD simulations were performed using LAMMPS (version: 19 Nov. 2024, Development) [22] to study Cl^- adsorption on Ni surfaces by comparing the grain boundary and interior regions. A symmetric tilt {210}[001] grain boundary was constructed using Atomsk 0.13.1 [23] by rotating two [001] grains $\pm 26.57°$. Cl ions were placed above the Ni surface, with two at the grain boundary and five in the grain interior to maintain charge neutrality. For the qualitative simulation, the interactions were modeled via a hybrid potential combining the embedded atom method (EAM) for Ni–Ni interactions [24] and customized Lennard–Jones (LJ) potentials for Ni–Cl (ε = 0.0025382 eV, σ = 2.300 Å) and Cl–Cl (ε = 0.00986272 eV, σ = 3.516 Å) interactions, with long-range Coulombic forces truncated at 10.0 Å and calculated using the particle–particle–mesh (PPPM) method. The Ni substrate was fixed, and simulations were performed at 300 K for 50,000 steps, that is, 25 ps, using a Nosé–Hoover thermostat. The adsorption energies were calculated from the stabilized potential energy of Cl ions in each region, and the results were visualized using OVITO 3.11.3 [25].

3. Results

Figure 1 shows the XRD patterns of the as-prepared Ni coatings at various current densities from 1 mA/cm^2 to 50 mA/cm^2. All the samples have two face-centered cubic (FCC) structures, i.e., electrodeposited Ni and the Cu substrate. The characteristic Ni peaks correspond to the $(111)_{FCC}$, $(200)_{FCC}$, and $(220)_{FCC}$ planes. Interestingly, these Ni peaks vary with the applied current density. This reflects the structural differences among the electrodeposited Ni coatings. At the lowest current density, the Ni peaks are broad and weak. This suggests the presence of fine grains in the samples. With increasing current density, the Ni peaks become relatively sharp and intense. This indicates a larger grain size in the sample. At current densities ranging from 1 mA/cm^2 to 20 mA/cm^2, the samples exhibit a preferred orientation along the (111) plane, corresponding to the plane with the lowest surface energy in FCC metals. At the largest current density of 50 mA/cm^2, the samples exhibit a strong (220) orientation in FCC crystal growth. This aligns with our previous work [16], which demonstrates that high current densities produce strongly oriented coatings. The shift from the close-packed (111) plane to the (220) plane increases the surface energy, contributing to reduced corrosion resistance. Cu diffraction peaks remain stable as the current density increases from 1 mA/cm^2 to 20 mA/cm^2, and the intensity decreases at a current density of 50 mA/cm^2, indicating a thicker Ni coating.

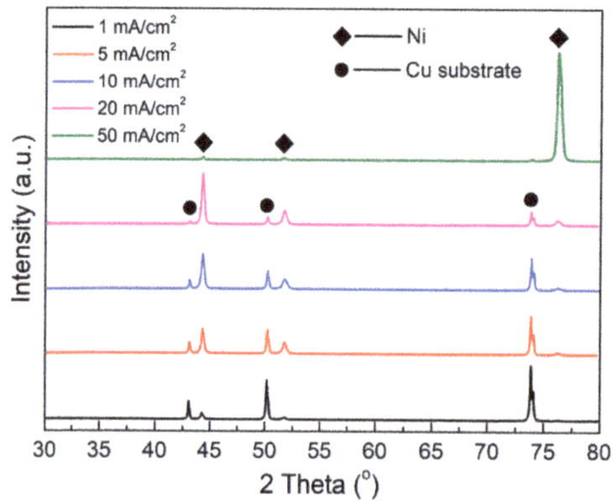

Figure 1. XRD patterns of Ni coatings with various current densities from 1 mA/cm^2 to 50 mA/cm^2.

Figure 2 shows the surface morphologies of the electrodeposited Ni coatings obtained at different current densities from 1 mA/cm^2 to 50 mA/cm^2. The grain size increases with the current density, which is consistent with the XRD results. Notably, the grain size heterogeneity varies with the tuned current density. At the lowest current density, the grains are uniformly small. At a current density of 5 mA/cm^2, the grains become heterogeneous, combining small and large grains, suggesting a transition in the nucleation–growth balance. The heterogeneity peaks at a current density of 10 mA/cm^2, with small and large grains coexisting. At a current density of 20 mA/cm^2, heterogeneity remains, but larger grains dominate. At the largest current density of 50 mA/cm^2, the grains are overgrown, and the heterogeneity is almost lost. This corresponds to the texturing observed in the XRD results at the highest current density. This heterogeneity reflects the interplay between nucleation and growth during electrodeposition [26]. The peak heterogeneity in the sample with a current density of 10 mA/cm^2 arises from the balance between nucleation and growth. A sufficiently high deposition rate [27] promotes grain growth, preferentially enlarging some grains, while numerous active nucleation sites sustain smaller grains. At lower current densities, nucleation dominates growth. With further increases in the current density, the loss of heterogeneity indicates that growth gradually dominates.

Polarization curves were measured to evaluate the corrosion resistance of the Ni coatings in a 3.5 wt.% NaCl solution (Figure 3a). The self-corrosion current density, I_{corr}, and self-corrosion potential, E_{corr}, were fitted based on the polarization curves and are listed in Figure 3b. The samples exhibit different corrosion behaviors in 3.5 wt.% NaCl solution. The self-corrosion current density is the lowest in the sample with a current density of 10 mA/cm^2 (4.48 μA/cm^2) and the highest in the sample with a current density of 1 mA/cm^2 (24.61 μA/cm^2). The self-corrosion potential is most positive for the samples with current densities of 5 mA/cm^2, 10 mA/cm^2, and 20 mA/cm^2 (−0.38 V, −0.37 V, and

−0.39 V, respectively). For the sample with the largest current density, the self-corrosion current density increases to 16.47 μA/cm^2, whereas the self-corrosion potential becomes most negative, i.e., −0.68 V. This indicates a reduced corrosion resistance and a change in the oxide kinetics. Beyond grain heterogeneity, the observed (220) texture contributes to this degradation, as its less close-packed structure increases the surface energy and enhances Cl^- adsorption compared to the (111)-oriented surfaces at lower current densities.

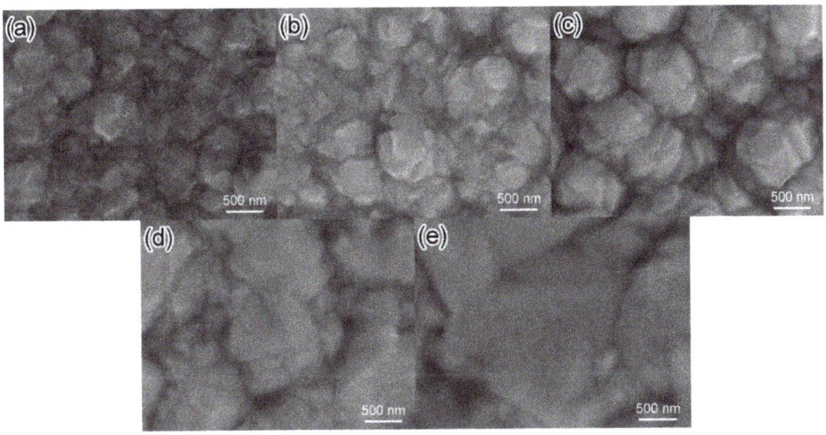

Figure 2. SEM images of Ni coatings at various current densities. (**a**) 1 mA/cm^2. (**b**) 5 mA/cm^2. (**c**) 10 mA/cm^2. (**d**) 20 mA/cm^2. (**e**) 50 mA/cm^2.

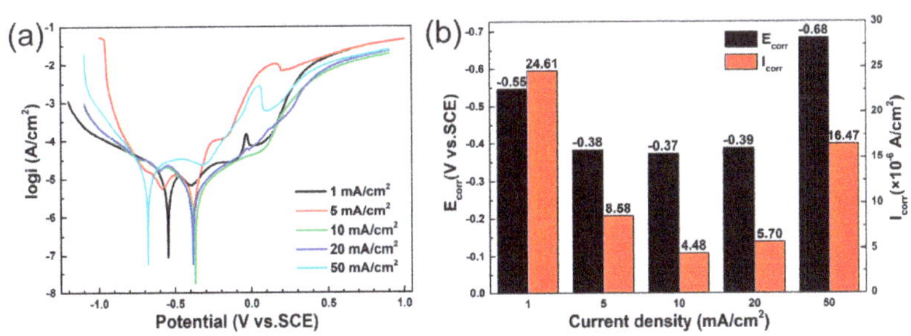

Figure 3. (**a**) Polarization curves of Ni coatings with various current densities from 1 mA/cm^2 to 50 mA/cm^2. (**b**) Fitted self-corrosion current density and potential.

Figure 4 shows the EIS Nyquist plots of the electrodeposited Ni coatings with the fitting curves. The semicircles in the curves show the charge transfer and mass transport processes at the interface between the Ni coating and the 3.5 wt% NaCl solution. The Nyquist plots with current densities from 5 mA/cm^2 to 50 mA/cm^2 show much larger semicircles than those with a current density of 1 mA/cm^2. An equivalent circuit, $R_1(Q_1R_2)(Q_2R_3)$ [28], was selected to fit the EIS data, where R_1 is the solution resistance, Q_1 and Q_2 are constant phase elements, and R_2 and R_3 represent the charge transfer (Faraday) processes. This circuit was chosen to capture the two-time constants observed in the Nyquist plots, consistent with the electrochemical behavior of the Ni coatings in 3.5 wt.% NaCl solution.

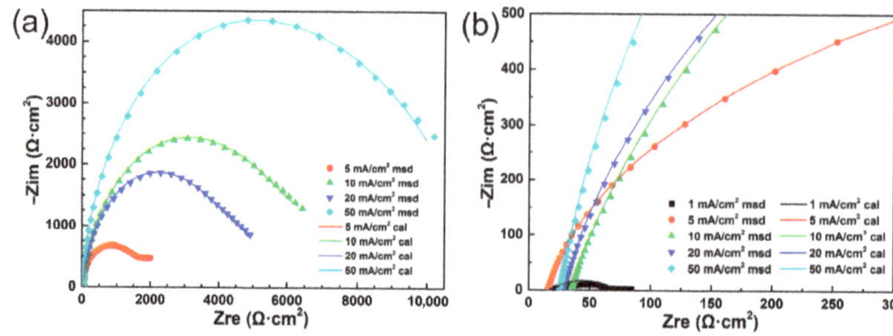

Figure 4. (**a**) EIS Nyquist plots of Ni coatings with various current densities from 1 mA/cm^2 to 50 mA/cm^2 corresponding to the fitting curves. (**b**) Enlarged part of the Figure shows the low-magnitude data.

The calculated values of the parameters are listed in Table 1. R_2 is the charge-transfer resistance at the interface between the Ni coatings and the 3.5 wt% NaCl solution, while R_3 represents a deeper charge-transfer process. The maximum R_2 is 4586 $\Omega{\cdot}cm^2$ in the sample with a current density of 10 mA/cm^2. This indicates a strong charge-transfer resistance. The minimum value is 23.6 $\Omega{\cdot}cm^2$ in the sample with a current density of 1 mA/cm^2. This suggests an enhanced ion exchange across the reactive surface. Conversely, the maximum R_3 is 7739 $\Omega{\cdot}cm^2$ in the sample with a current density of 50 mA/cm^2. This implies an additional mechanism, possibly the formation of a thicker corrosion product layer during the secondary charge transfer. However, the lower R_2 and negative self-corrosion potential of the sample indicate that this layer offers limited corrosion protection.

Table 1. Calculated parameters of the equivalent circuits of the EIS results.

Current Density (mA/cm^2)	R_1 ($\Omega{\cdot}cm^2$)	Q_1-Y_0 ($\Omega^{-1}{\cdot}cm^2{\cdot}S^n$)	Q_1-n	R_2 ($\Omega{\cdot}cm^2$)	Q_2-Y_0 ($\Omega^{-1}{\cdot}cm^2{\cdot}S^n$)	Q_2-n	R_3 ($\Omega{\cdot}cm^2$)
1	15.62	0.00193	1	23.6	0.00855	0.4228	36.96
5	15.85	8.716×10^{-4}	0.7893	1107	2.995×10^{-5}	0.9158	1410
10	34.54	2.641×10^{-5}	0.905	4586	2.983×10^{-4}	0.8799	2473
20	28.54	2.866×10^{-5}	0.9312	3554	4.238×10^{-4}	0.8431	1773
50	25.82	2.689×10^{-4}	0.9792	3295	2.713×10^{-5}	0.9502	7739

The corrosion products of the electrodeposited Ni coatings prepared at a current density of 10 mA/cm^2 were characterized using TEM after Tafel polarization. In Figure 5a, the bright-field image shows the petal-like nanostructures. In Figure 5b,c, HAADF-EDS mapping reveals the presence of Ni. As shown in Figure 5d, the HRTEM results confirm the corrosion products. The fast Fourier transform (FFT) pattern inset is obtained from the area in the red box with Z = $[001]_{Ni}$. The inset of the inverse FFT (IFFT) results shows that the d-spacing of $(011)_{Ni}$ is 0.26 nm. Interestingly, Figure 5e shows NiOOH at the periphery of Ni. Compared with Ni, NiOOH exhibits a larger d-spacing of 0.59 nm for the (0001) plane [29]. NiOOH facilitates the formation of a passivation layer [30,31]. In 3.5 wt.% NaCl medium, it provides initial passivation [32], reducing coating dissolution.

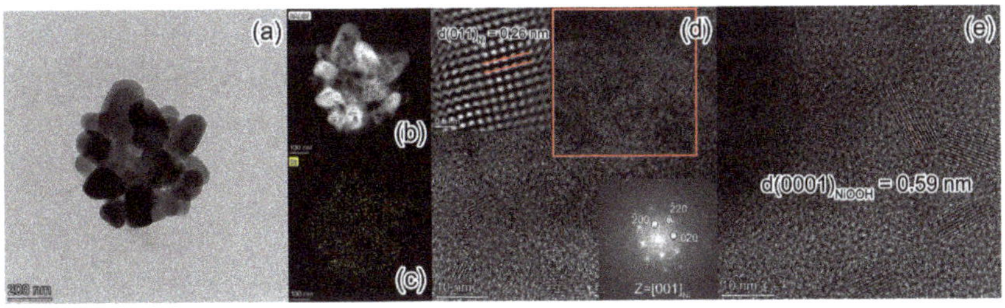

Figure 5. TEM results of the sample at a current density of 10 mA/cm^2 after polarization testing. (**a**) Bright-field image. (**b**) HAADF image. (**c**) EDS mapping of Ni. (**d**,**e**) HRTEM images of the corrosion products.

Figure 6 shows the TEM results of the corrosion products of the Ni coatings at a current density of 50 mA/cm^2. In Figure 6a, the bright-field image shows irregular, loosely packed nanostructures, which differ from the compact clusters of the sample with a current density of 10 mA/cm^2. The inset selected-area electron diffraction (SAED) patterns in Figure 6a display diffuse rings. This confirms the amorphous nature [33] of the corrosion products. Porous amorphous corrosion products reduce protective efficacy by facilitating Cl^- penetration [34], likely due to their higher permeability and reduced barrier properties. A crystalline oxide inner layer is typically covered by an outer NiOOH/$Ni(OH)_2$ layer at high potentials [35]. Under high current density, elevated overpotential, or prolonged exposure, the NiO layer grows rapidly and disorderedly due to irreversible oxidation reactions [35]. This disordered growth produces loosely packed structures, increasing susceptibility to Cl^- adsorption and subsequent corrosion. This property, coupled with hydration, remarkably increases the surface area exposed to corrosive species, promoting localized corrosion, such as pitting, and increasing corrosion rates, making it less protective than NiOOH. Thus, these amorphous corrosion products indicate a significant corrosion severity. This transition likely results from the elevated anodic overpotential for the sample with the largest current density, which accelerates Ni oxidation and promotes the rapid growth of a disordered structure [35], while the lower overpotential for the samples with lower current densities supports the formation of ordered NiOOH. Correspondingly, the sample exhibited reduced corrosion resistance, as indicated by a high I_{corr} of 16.47 μA/cm^2 and E_{corr} of −0.68 V. The highest R_3, 7739 Ω·cm^2, was also observed in the EIS results. This can be interpreted as resistance associated with a deeper, diffusion-limited process within the amorphous corrosion product, such as $NiO{\cdot}nH_2O$. Figure 6b,c confirms the presence of Ni through HAADF-EDS mapping. In Figure 6d, no lattice fringes are observed in the HRTEM images. Corresponding to the red boxed region in Figure 6d, the FFT inset shows diffuse and broad rings. These results confirm the non-crystalline character, i.e., the lack of long-range order.

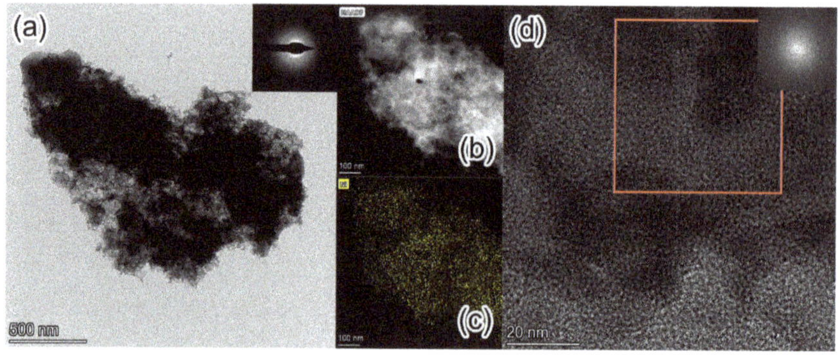

Figure 6. TEM results of the sample at a current density of 50 mA/cm^2 after polarization testing. (**a**) Bright-field image. (**b**) HAADF image. (**c**) EDS mapping of Ni. (**d**) HRTEM results of the corrosion products.

4. Discussion

The corrosion resistance of the Ni-coated electrode, which was electrodeposited in 3.5 wt.% NaCl solution is fundamentally related to their microstructure. In this study, the microstructure was customized by tuning the applied current density during electrodeposition (from 1 to 50 mA/cm^2). The prepared heterostructures remarkably enhanced corrosion resistance by leveraging extended grain diversity to mitigate corrosion. MD simulations were carried out to understand the adsorption mechanism of Cl^- on the Ni surface in the earliest stage of corrosion (Figure 7). MD simulations revealed stronger Cl^- adsorption at the grain boundaries (−4.36 eV) than at the grain interiors (−1.74 eV), offering additional insights that complement the density functional theory (DFT) calculations [36]. The deeper energy well at the boundary (2.62 eV difference) shows enhanced electrostatic and van der Waals interactions [37], indicating that the grain boundaries are preferential sites for attack before passivation layers such as NiOOH or $NiO{\cdot}nH_2O$ are formed. Thus, grain boundaries serve as initial adsorption sites before passivation layers such as NiOOH or $NiO{\cdot}nH_2O$. Furthermore, the texture evolution from (111) to (220) at 50 mA/cm^2 amplifies this degradation, highlighting the critical role of the grain structure and crystal orientation in governing the corrosion behavior.

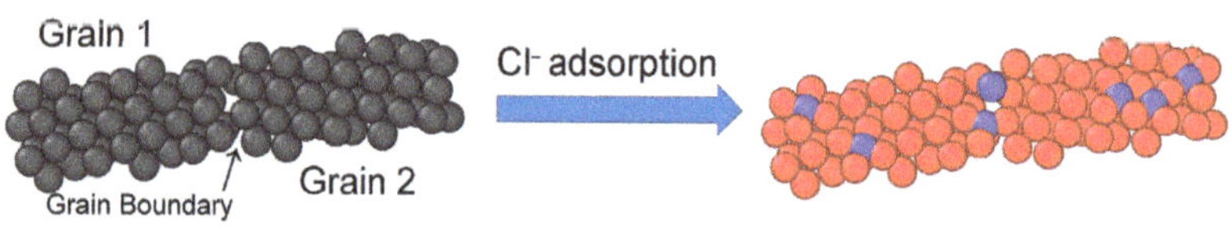

Figure 7. MD simulation results for Cl^- adsorption on the Ni surfaces, including the grain boundary and grain interior regions.

In the sample with the smallest current density, the high density of nanoscale boundaries amplifies this weakness, as reflected in its electrochemical properties. In such nanocrystalline materials, triple junctions, i.e., points where three-grain boundaries intersect, exacerbate this effect below the critical grain size. Naturally, these junctions increase the defect density and provide pathways for corrosive species, such as Cl^-, due to the segregation of impurities or alloying elements at the triple junction [38]. Additionally, the triple-junction diffusion coefficient for Ni (at d = 70 nm) exceeds that of the grain boundaries (3.5×10^{-9} m^2/s vs. 2.8×10^{-26} m^2/s at 298 K) [39]. This enhances atomic mobility and potentially affects passive film formation during corrosion. The triple-junction volume fraction V_{tj} is estimated as follows [40],

$$V_{tj} = V_{ic} - V_{gb} \tag{1}$$

where V_{ic} and V_{gb} are the intercrystalline and grain boundary volume fractions, respectively [40],

$$V_{ic} = 1 - \left(\frac{d-\delta}{d}\right)^3 \tag{2}$$

$$V_{gb} = \frac{\left[3\delta(d-\delta)^2\right]}{d^3} \tag{3}$$

where d is the grain size and where δ is the grain boundary thickness. This indicates a significant triple-junction network in the nanocrystalline material that can enhance atomic mobility compared to coarser structures [39]. In contrast, the prepared heterogeneous Ni coatings render this susceptibility a favorable mechanism. Compared to uniform nanograins, the mixture of small and large nanograins disperses Cl^- adsorption and mitigates localized corrosion, which may promote the formation of a NiOOH protective layer.

Synergy spans multiple scales in different heterogeneous materials, such as Fe–Cr–Ni alloys [41], AA 7075 Al alloys [42], and CoCrFeMnNi high-entropy alloys [43]. As corrosion progresses beyond initiation, nanoscale or ultrafine regions accelerate passivation kinetics, while microscale or coarse regions enhance durability and reduce oxidation and pitting susceptibility across diverse material systems and conditions [41–43]. The role of structural heterogeneity in corrosion resistance can be understood using a mechanical analogy. Similar to the way smaller nanograins impede dislocation motion to enhance strength, they also promote passivation. Larger grains provide structural stability during corrosion, similar to their role in maintaining ductility in heterogeneous materials [9]. This strategy exemplifies the potential of compositional plainification to achieve superior performance, such as corrosion resistance, through structural tuning rather than compositional complexity [3]. This approach optimizes material performance via tailored interface engineering with simpler compositions and pure elements. Traditional alloying, as seen in steel, superalloys, and high-entropy alloys strengthened with elements like Cr [44], often compromises manufacturing sustainability and complicates recycling [45]. As demonstrated with pure Ni coatings, compositional plainification enables stable nanoscale interfaces to enhance corrosion resistance without alloying. The heterogeneous structure can also enhance the mechanical properties, such as hardness, wear resistance, and fatigue resistance [9]. This work highlights the role of structural heterogeneity in corrosion properties, supporting a sustainable approach to high-performance coating design.

5. Conclusions

This study introduces a one-step, scalable electrodeposition approach to fabricate Ni coatings with tunable nanocrystalline heterostructures, and their corrosion resistance can be optimized by varying the current density from 1 to 50 mA/cm^2. The experimental results demonstrate that the heterostructure formed at 10 mA/cm^2, characterized by co-existing small and large grains, exhibits superior corrosion resistance in a 3.5 wt.% NaCl solution, with a self-corrosion current density as low as 4.48 μA/cm^2, outperforming other conditions. Molecular dynamics simulations revealed that stronger Cl^- adsorption at grain boundaries than at grain interiors initiates corrosion, whereas the heterostructure mitigates this effect by dispersing the adsorption sites. High-resolution transmission electron microscopy further revealed that as the current density increased from 10 mA/cm^2 to

50 mA/cm^2, the corrosion product shifted from a crystalline NiOOH structure to an amorphous structure, reducing the passivation efficacy. This structural optimization strategy avoids the complexity of alloying, leverages the heterogeneous microstructure of pure Ni to enhance performance, and offers a cost-effective and manufacturable coating option for offshore energy and chemical engineering. By integrating electrochemical testing, microstructural characterization, and simulations, this work elucidates the multiscale synergistic mechanisms underpinning enhanced corrosion resistance, providing clear guidance for designing high-performance metallic coatings.

Author Contributions: W.H.: Conceptualization, Methodology, Investigation, Visualization, Writing—original draft, Writing—review and editing, and Supervision. Z.Z.: Investigation, Formal analysis, Visualization, Writing—original draft. X.H.: Investigation. Y.W.: Investigation. S.W.: Investigation. X.L.: Investigation. S.L.: Investigation. S.Z.: Investigation. F.F.: Writing—review and editing. J.J.: Writing—review and editing. All authors have read and agreed to the published version of the manuscript.

Funding: This work was supported by the Natural Science Foundation of Jiangsu Province, China (BK20220428) and Jiangsu Students' Platform for Innovation and Entrepreneurship Training Program (202210298091Y). W.H. acknowledges support from the European Union Horizon 2020 research and innovation program under grant agreement no. 857470, the European Regional Development Fund via the Foundation for Polish Science International Research Agenda PLUS program grant no. MAB PLUS/2018/8. This publication was partly created within the framework of the project of the Minister of Science and Higher Education, Support for the activities of Centres of Excellence established in Poland under Horizon 2020, under contract no. MEiN/2023/DIR/3795.

Institutional Review Board Statement: Not applicable.

Informed Consent Statement: Not applicable.

Data Availability Statement: Data will be made available upon request.

Acknowledgments: W.H. acknowledges the computational resources provided by the High-Performance Cluster at the National Centre for Nuclear Research in Poland.

Conflicts of Interest: The authors declare no conflicts of interest.

References

1. Wang, Y.; Zhou, X.; Lu, Y.; Li, X. Enhanced mechanical properties in bulk nanograined Ni with high-density fivefold twins. *Small* **2025**, *21*, 2410202. [CrossRef]
2. Huang, C.; Zhang, Z.; Yan, J.; Sun, L.; Wang, J. Enhancing wear and corrosion resistance of electroless Ni-P coatings in CO_2-saturated NaCl solution through polytetrafluoroethylene incorporation. *Corros. Sci.* **2024**, *226*, 111620. [CrossRef]
3. Li, X.; Lu, K. Improving sustainability with simpler alloys. *Science* **2019**, *364*, 733–734. [CrossRef]
4. Liu, D.; Wang, D.; Hong, T.; Wang, Z.; Wang, Y.; Qin, Y.; Su, L.; Yang, T.; Gao, X.; Ge, Z.; et al. Lattice plainification advances highly effective SnSe crystalline thermoelectrics. *Science* **2023**, *380*, 841–846. [CrossRef]
5. Naghdi, A.; Domínguez-Gutiérrez, F.J.; Huo, W.Y.; Karimi, K.; Papanikolaou, S. Dynamic nanoindentation and short-range order in equiatomic NiCoCr medium-entropy alloy lead to novel density wave ordering. *Phys. Rev. Lett.* **2024**, *132*, 116101. [CrossRef]
6. Zhao, S.; Zhang, R.; Yu, Q.; Ell, J.; Ritchie, R.O.; Minor, A.M. Cryoforged nanotwinned titanium with ultrahigh strength and ductility. *Science* **2021**, *373*, 1363–1368. [CrossRef]
7. Li, Y.; Yuan, G.; Li, L.; Kang, J.; Yan, F.; Du, P.; Raabe, D.; Wang, G. Ductile 2-GPa steels with hierarchical substructure. *Science* **2023**, *379*, 168–173. [CrossRef]
8. Li, X.Y.; Zhou, X.; Lu, K. Rapid heating induced ultrahigh stability of nanograined copper. *Sci. Adv.* **2020**, *6*, eaaz8003. [CrossRef]
9. Zhu, Y.; Wu, X. Heterostructured materials. *Prog. Mater. Sci.* **2023**, *131*, 101019. [CrossRef]
10. Wang, Y.; Zhu, Y.; Yu, Z.; Zhao, J.; Wei, Y. Hetero-zone boundary affected region: A primary microstructural factor controlling extra work hardening in heterostructure. *Acta Mater.* **2022**, *241*, 118395. [CrossRef]
11. Kong, Y.; Peng, K.; Huang, H. Highly controllable additive manufacturing of heterostructured nickel-based composites. *Int. J. Mach. Tool. Manuf.* **2024**, *195*, 104112. [CrossRef]

12. Peng, B.; Jin, J.; Liu, Y.; Lu, C.; Li, L.; Yan, M. Towards peculiar corrosion behavior of multi-main-phase Nd-Ce-Y-Fe-B permanent material with heterogeneous microstructure. *Corros. Sci.* **2020**, *177*, 108972. [CrossRef]
13. Zhao, R.; Wang, H.; Du, H.; Yang, Y.; Gao, Z.; Qie, L.; Huang, Y. Lanthanum nitrate as aqueous electrolyte additive for favourable zinc metal electrodeposition. *Nat. Commun.* **2022**, *13*, 3252. [CrossRef]
14. Ye, C.; Jin, H.; Shan, J.; Jiao, Y.; Li, H.; Gu, Q.; Davey, K.; Wang, H.; Qiao, S.-Z. A Mo_5N_6 electrocatalyst for efficient Na_2S electrodeposition in room-temperature sodium-sulfur batteries. *Nat. Commun.* **2021**, *12*, 7195. [CrossRef]
15. Li, Y.; Cai, X.; Zhang, G.; Xu, C.; Gao, W.; An, M. Optimization of electrodeposition nanocrytalline Ni-Fe alloy coatings for the replacement of Ni coatings. *J. Alloy. Comp.* **2022**, *903*, 163761. [CrossRef]
16. Huo, W.; Wang, S.; Fang, F.; Tan, S.; Kurpaska, Ł.; Xie, Z.; Kim, H.S.; Jiang, J. Microstructure and corrosion resistance of highly <111> oriented electrodeposited CoNiFe medium-entropy alloy films. *J. Mater. Res. Technol.* **2022**, *20*, 1677–1684. [CrossRef]
17. Li, S.; Li, H.; Zhai, Z.; Cao, X.; Liu, D.; Jiang, J. Corrosion resistance and tribological behavior of FeCoCrNi@GO/Ni high entropy alloy-based composite coatings prepared by electrodeposition. *Surf. Coating. Technol.* **2024**, *477*, 130379. [CrossRef]
18. Dong, M.; Liu, P.; Wang, C.; Wang, Y.; Tang, X.; He, M.; Liu, J. Microstructure and properties of FeCoNiCr and FeCoNiCrW high entropy alloy coatings by electrodeposition. *Intermetallics* **2024**, *175*, 108492. [CrossRef]
19. Wang, Y.; Guan, L.; He, Z.; Zhang, S.; Singh, H.; Hayat, M.D.; Yao, C. Influence of pretreatments on physicochemical properties of Ni-P coatings electrodeposited on aluminum alloy. *Mater. Des.* **2021**, *197*, 109233. [CrossRef]
20. Daneshnia, A.; Raeissi, K.; Salehikahrizsangi, P. Rapid one-step electrodeposition of robust superhydrophobic and oleophobic Ni coating with anti-corrosion and self-cleaning properties. *Surf. Coating. Technol.* **2022**, *450*, 129007. [CrossRef]
21. Wang, Y.; Zhang, G.; He, Z.; Chen, J.; Gao, W.; Cao, P. Superhydrophobic Ni nanocone surface prepared by electrodeposition and its overall performance. *Surf. Coating. Technol.* **2023**, *464*, 129548. [CrossRef]
22. Thompson, A.P.; Aktulga, H.M.; Berger, R.; Bolintineanu, D.S.; Brown, W.M.; Crozier, P.S.; in't Veld, P.J.; Kohlmeyer, A.; Moore, S.G.; Nguyen, T.D.; et al. LAMMPS—A flexible simulation tool for particle-based materials modeling at the atomic, meso, and continuum scales. *Comp. Phys. Comm.* **2022**, *271*, 108171. [CrossRef]
23. Hirel, P. Atomsk: A tool for manipulating and converting atomic data files. *Comput. Phys. Comm.* **2015**, *197*, 212–219. [CrossRef]
24. Foiles, S.M.; Baskes, M.I.; Daw, M.S. Embedded-atom-method functions for the fcc metals Cu, Ag, Au, Ni, Pd, Pt, and their alloys. *Phys. Rev. B* **1986**, *33*, 7983–7991. [CrossRef]
25. Stukowski, A. Visualization and analysis of atomistic simulation data with OVITO—The Open Visualization Tool. Modelling Simul. *Mater. Sci. Eng.* **2010**, *18*, 015012. [CrossRef]
26. Lin, J.; Kilani, M.; Baharfar, M.; Wang, R.; Mao, G. Understanding the nanoscale phenomena of nucleation and crystal growth in electrodeposition. *Nanoscale* **2024**, *16*, 19564–19588. [CrossRef]
27. Chen, L.; Zhang, G.; Zhou, G.; Xiang, C.; Miao, X.; Liu, L.; An, X.; Lan, H.; Liu, H. In situ visual observation of surface energy-controlled heterogeneous nucleation of metal nanocrystals. *Small* **2024**, *20*, 2401674. [CrossRef]
28. Zhou, J.; Cheng, Y.; Wan, Y.; Chen, H.; Wang, Y.; Ma, K.; Yang, J. Enhancement mechanisms of self-lubricating Ti_3SiC_2 ceramic doping in CoCrFeNi high-entropy alloy via high-speed laser cladding: Tribology and electrochemical corrosion. *Surf. Coating. Technol.* **2024**, *480*, 130554. [CrossRef]
29. Dai, L.; Fang, C.; Yao, F.; Zhang, X.; Xu, X.; Han, S.; Deng, J.; Zhu, J.; Sun, J. Thickness-dependent β/γ-NiOOH transformation of Ni-MOFs in oxygen evolution reaction. *Appl. Surf. Sci.* **2023**, *623*, 156991. [CrossRef]
30. Qin, L.; Lian, J.; Jiang, Q. Effect of grain size on corrosion behavior of electrodeposited bulk nanocrystalline Ni. *Trans. Nonferr. Metal. Soc. China* **2010**, *20*, 82–89. [CrossRef]
31. Hall, D.S.; Lockwood, D.J.; Bock, C.; MacDougall, B.R. Nickel hydroxides and related materials: A review of their structures, synthesis and properties. *Proc. R. Soc. A Math. Phys. Eng. Sci.* **2015**, *471*, 20140792. [CrossRef] [PubMed]
32. El-Tantawy, Y.A.; Al-Kharafi, F.M. Role of Cl− in breakdown of Ni passivity in aqueous NaOH solutions. *Electrochim. Acta* **1982**, *27*, 691–699. [CrossRef]
33. Wan, C.; Zhang, Z.; Dong, J.; Xu, M.; Pu, H.; Baumann, D.; Lin, Z.; Wang, S.; Huang, J.; Shah, A.H.; et al. Amorphous nickel hydroxide shell tailors local chemical environment on platinum surface for alkaline hydrogen evolution reaction. *Nat. Mater.* **2023**, *22*, 1022–1029. [CrossRef] [PubMed]
34. Marcus, P.; Maurice, V. Atomic level characterization in corrosion studies. *Philos. T. R. Soc. A* **2017**, *375*, 20160414. [CrossRef]
35. Davies, D.E.; Barker, W. Influence of pH on corrosion and passivation of nickel. *Corrosion* **1964**, *20*, 47t–53t. [CrossRef]
36. Yang, S.; Liang, G.; Huang, Y.; Hao, X.; Zhao, J.; Lv, M. Adsorption structure and properties of Ni/Fe electrodeposition interface: A DFT study. *Model. Simul. Mater. Sci. Eng.* **2024**, *32*, 055024. [CrossRef]
37. Zhang, P.; Yang, Y.; Duan, X.; Liu, Y.; Wang, S. Density functional theory calculations for insight into the heterocatalyst reactivity and mechanism in persulfate-based advanced oxidation reactions. *ACS Catal.* **2021**, *11*, 11129–11159. [CrossRef]
38. Wasekar, N.P. The influence of grain size and triple junctions on corrosion behavior of nanocrystalline Ni and Ni-W alloy. *Scr. Mater.* **2022**, *213*, 114604. [CrossRef]

39. Chen, Y.; Schuh, C.A. Contribution of triple junctions to the diffusion anomaly in nanocrystalline materials. *Scr. Mater.* **2007**, *57*, 253–256. [CrossRef]
40. Palumbo, G.; Thrope, S.J.; Aust, K.T. On the contribution of triple junctions to the structure and properties of nanocrystalline materials. *Scr. Metall. Mater.* **1990**, *24*, 1347–1350. [CrossRef]
41. Mahesh, B.V.; Raman, R.K.S.; Koch, C.C. Bimodal grain size distribution: An effective approach for improving the mechanical and corrosion properties of Fe–Cr–Ni alloys. *J. Mater. Sci.* **2012**, *47*, 7735–7743. [CrossRef]
42. Tian, W.; Li, S.; Wang, B.; Liu, J.; Yu, M. Pitting corrosion of naturally aged AA 7075 aluminum alloys with bimodal grain size. *Corros. Sci.* **2016**, *113*, 1–16. [CrossRef]
43. Wang, J.; Zhang, Z.; Dai, H.; Fujiwara, H.; Chen, X.; Ameyama, K. Enhanced corrosion resistance of CoCrFeMnNi high entropy alloy using heterogeneous structure design. *Corros. Sci.* **2022**, *209*, 110761. [CrossRef]
44. Li, Y.; Olejarz, A.; Kurpaska, Ł.; Lu, E.; Alava, M.J.; Kim, H.S.; Huo, W. Designing cobalt-free face-centered cubic high-entropy alloys: A strategy using d-orbital energy level. *Int. J. Refract. Metal. Hard Mater.* **2024**, *124*, 106834. [CrossRef]
45. Huo, W.; Wang, S.; Zhang, X.; Ren, K.; Tan, S.; Fang, F.; Xie, Z.; Jiang, J. A strategy to improve the performance of TiO_2 nanotube array film photocatalysts by magnetron-sputtered amorphous $BiFeO_3$. *Vacuum* **2022**, *202*, 111135. [CrossRef]

Disclaimer/Publisher's Note: The statements, opinions and data contained in all publications are solely those of the individual author(s) and contributor(s) and not of MDPI and/or the editor(s). MDPI and/or the editor(s) disclaim responsibility for any injury to people or property resulting from any ideas, methods, instructions or products referred to in the content.

coatings

MDPI

Article

Improvement in the Corrosion and Wear Resistance of ZrO_2-Ag Coatings on 316LVM Stainless Steel Under Tribocorrosive Conditions

Willian Aperador [1] and Giovany Orozco-Hernández [2,*]

[1] Department of Engineering, Universidad Militar Nueva Granada, Bogotá 110111, Colombia; william.aperador@unimilitar.edu.co
[2] Postgraduate Department, Universidad ECCI, Bogotá 111311, Colombia
* Correspondence: gorozcoh@ecci.edu.co

Abstract

This study investigates the development of silver (Ag)-doped zirconia (ZrO_2) coatings deposited on 316LVM stainless steel via the unbalanced magnetron sputtering technique. The oxygen content in the Ar/O_2 gas mixture was systematically varied (12.5%, 25%, 37.5%, and 50%) to assess its influence on the resulting coating properties. In response to the growing demand for biomedical implants with improved durability and biocompatibility, the objective was to develop coatings that enhance both wear and corrosion resistance in physiological environments. The effects of silver incorporation and oxygen concentration on the structural, tribological, and electrochemical behavior of the coatings were systematically analyzed. X-ray diffraction (XRD) was employed to identify crystalline phases, while atomic force microscopy (AFM) was used to characterize surface topography prior to wear testing. Wear resistance was evaluated using a ball-on-plane tribometer under simulated prosthetic motion, applying a 5 N load with a bone pin as the counter body. Corrosion resistance was assessed through electrochemical impedance spectroscopy (EIS) in a physiological solution. Additionally, tribocorrosive performance was investigated by coupling tribological and electrochemical tests in Ringer's lactate solution, simulating dynamic in vivo contact conditions. The results demonstrate that Ag doping, combined with increased oxygen content in the sputtering atmosphere, significantly improves both wear and corrosion resistance. Notably, the ZrO_2-Ag coating deposited with 50% O_2 exhibited the lowest wear volume (0.086 mm^3) and a minimum coefficient of friction (0.0043) under a 5 N load. This same coating also displayed superior electrochemical performance, with the highest charge transfer resistance (38.83 $k\Omega \cdot cm^2$) and the lowest corrosion current density (3.32×10^{-8} A/cm^2). These findings confirm the high structural integrity and outstanding tribocorrosive behavior of the coating, highlighting its potential for application in biomedical implant technology.

Keywords: thin films; sputtering; 316LVM; electrochemical properties; microstructure; biocompatibility; hardness

1. Introduction

316LVM stainless steel is one of the most widely used materials for medical implants due to its excellent corrosion resistance, good biocompatibility, and favorable mechanical properties [1–4]. However, despite these advantages, implants made from 316LVM face several long-term challenges within the human body [5–7]. These include the formation of

 https://doi.org/10.3390/coatings15080862

unstable passive layers, the release of potentially toxic metal ions, and inadequate osseointegration. Such issues can trigger adverse biological responses, including inflammation, infection, and implant rejection, thereby compromising implant longevity and patient outcomes [8–10].

Zirconium oxide (ZrO_2) has garnered significant attention for biomedical applications because of its high mechanical strength, chemical inertness, and superior biocompatibility [11]. These attributes make ZrO_2 an attractive candidate for protective coatings on metallic implants, where resistance to wear, corrosion, and degradation in physiological environments is essential [12]. Nonetheless, long-term exposure to bodily fluids and mechanical stress may impair its adhesion and mechanical integrity, limiting long-term functionality.

To address these limitations, the incorporation of silver (Ag) nanoparticles into ZrO_2 matrices has been explored as a strategy to enhance both the structural integrity and functional performance of these coatings [13]. Silver exhibits broad-spectrum antimicrobial activity, making it effective in reducing the risk of implant-associated infections, while also contributing to improved corrosion and wear resistance. Despite promising outcomes, the influence of Ag doping on the mechanical, structural, and tribocorrosive behavior of ZrO_2 coatings remains an active area of investigation [14–16].

Silver-doped zirconium oxide (ZrO_2-Ag) coatings represent a class of advanced ceramic nanocomposites with significant potential for biomedical applications. The addition of Ag nanoparticles has been shown to enhance the structural integrity, surface characteristics, and functional durability of ZrO_2-based coatings. These improvements are critical for maintaining coating performance under dynamic physiological conditions, where conventional materials often suffer from reduced wear resistance and compromised adhesion [14].

To overcome the limitations of the materials used in medical implants, the use of zirconium oxide (ZrO_2) has been explored. ZrO_2 is a ceramic material that excels in biomedical applications due to its high mechanical strength, biocompatibility, and stability [16–18]. However, its adherence and wear resistance may be compromised under physiological conditions. Doping ZrO_2 with silver nanoparticles (Ag NPs) has proven to be an effective strategy for enhancing these properties, increasing corrosion resistance and enhancing the antibacterial characteristics of the material [19]. This optimizes the durability of the implants and prevents infections in the surgical environment. These advancements have led to ceramic nanocomposites such as ZrO_2-Ag being considered promising options in the development of advanced coatings for medical implants, as they improve biocompatibility and corrosion resistance and enhance adherence, an essential aspect for ensuring implant longevity without generating toxicity in surrounding tissues [20,21].

Despite these advances, several challenges remain in translating ZrO2-Ag coatings to clinical use. Key concerns include ensuring long-term adhesion to metallic substrates, maintaining mechanical performance under tribological stress, and fully understanding the coatings' interactions with biological tissues [22]. In this context, the present study aims to evaluate the tribological and electrochemical properties of ZrO_2-Ag coatings deposited on 316LVM stainless steel substrates [23].

The use of advanced deposition techniques, such as magnetron sputtering, has enabled the production of homogeneous and adherent ceramic coatings with enhanced performance [24,25]. However, optimizing adhesion, hardness, and wear resistance under physiological and mechanical loading conditions remains a challenge [6]. This study addresses these issues by characterizing the performance of ZrO_2-Ag coatings under simulated tribocorrosion conditions.

The coatings were analyzed using X-ray diffraction (XRD) to identify phase composition and [26] atomic force microscopy (AFM) and assess surface morphology and triboelectrochemical testing to simulate fretting corrosion. Electrochemical impedance spectroscopy (EIS) was employed to evaluate corrosion resistance in physiological media [27,28]. By correlating the coating composition and deposition parameters with tribocorrosive performance, this study aims to advance the development of durable, multifunctional coatings for next-generation biomedical implants.

2. Materials and Methods

Silver-doped zirconium oxide (ZrO_2-Ag) thin films were deposited on 316LVM stainless steel substrates using the magnetron sputtering technique. Moreover, 316LVM, an austenitic stainless steel widely used in biomedical applications due to its excellent corrosion resistance and biocompatibility, served as the base material. Its chemical composition, determined by energy-dispersive spectroscopy (EDS), includes 16.4 at.% chromium (Cr), 69.9 at.% iron (Fe), 10.1 at.% nickel (Ni), 1.5 at.% manganese (Mn), and 2.1 at.% molybdenum (Mo). This alloy exhibits a yield strength of approximately 200 MPa, tensile strength ranging from 550 to 750 MPa, and an elongation of about 40%, combining mechanical strength with ductility. The sputtering chamber was evacuated to a base pressure of ~10^{-6} mbar, after which argon (Ar) gas was introduced at a flow rate of 25 SCCM (standard cubic centimeters per minute) and oxygen at 5 SCCM to initiate plasma formation. Deposition was performed at a working pressure of 3.0×10^{-2} mbar and room temperature, using a target power of 40 W. During coating, substrates were rotated at 6 rpm to ensure uniform film coverage.

For deposition, 316LVM stainless steel disks (diameter = 1.27 cm and thickness = 3 mm) and silicon wafers (Si (100)) were used. Silicon substrates were employed for coating thickness measurements. Prior to deposition, steel samples were polished using sandpaper followed by 1 μm and 0.3 μm alumina suspensions, achieving a surface roughness of 37 ± 2 nm.

ZrO_2-Ag coatings were deposited using RF magnetron sputtering at 13.56 MHz. A 4-inch Zr target with 99.9% purity was used, positioned 4.3 cm from the substrate. The pressure in the chamber was set at 2×10^{-2} mbar, with a cathode power of 80 W. The deposition conditions were maintained at 50 W power, 2.6×10^{-3} mbar working pressure, and a substrate temperature of 300 °C. The oxygen content in the Ar/O_2 gas mixture was varied (12.5%, 25%, 37.5%, and 50%) to evaluate its effect on coating properties.

The microstructure and nanocomposite formation of the ZrO_2-Ag coatings were analyzed using JEOL JEM 2100 transmission electron microscopy (TEM) at 200 kV, manufactured by JEOL Ltd., Tokyo, Japan. The coatings deposited on 316LVM stainless steel substrates were cross-sectioned for further characterization, transferred to a copper grid, and thinned with a precision microtome to below 100 nm. The imaging parameters were optimized to capture high-resolution structural details.

Crystallographic characterization was conducted via X-ray diffraction (XRD) using an EMPYREAN-PANalytical manufactured by PANalytical B.V., Almelo, The Netherlands. System with monochromatic Cu-Kα radiation (λ = 1.5406 Å) in Bragg–Brentano geometry. Spectra were acquired in a 2θ range of 10° to 90°, with a step size of 0.02° and a step time of 2 s. The crystalline phases were identified by comparing peak positions with entries from the Crystallography Open Database (COD-2022).

The surface morphology prior to wear testing was examined via atomic force microscopy (AFM, Asylum Research MFP-3D, manufactured by Asylum Research, an Oxford Instruments Company, Santa Barbara, CA, USA) in contact mode over a 45 μm × 45 μm area.

The wear resistance was evaluated using a linearly reciprocating ball-on-flat sliding tribometer (Nanovea T2000, manufactured by Nanovea, Irvine, CA, USA). The counter-body consisted of a 3 mm-diameter cylindrical pin fabricated via a micromachining tool, with a total distance of 6 cm from the bone. Although designed to simulate microscopic-scale contact conditions, the 5 N load applied corresponds to a significantly higher pressure than typical microscale applications (mN range). This experimental setup was chosen to replicate localized contact stresses encountered in biomedical implants. A sliding speed of 2 mm/s was applied over a total sliding distance of 888 m.

Electrochemical impedance spectroscopy (EIS) was conducted to assess corrosion resistance in a physiological medium using a Gamry's Interface 1010E potensiostat, Interface 1010E, manufactured by Gamry Instruments, Warminster, PA, USA. Ringer's physiological solution (9 g/L of NaCl, 0.4 g/L of $CaCl_2$, and 2.1 g/L of $NaHCO_3$) at 37 °C was used as the electrolyte. Measurements were performed with a 5 mV AC perturbation across a frequency range of 10 kHz to 0.001 Hz. The data were analyzed using Bode plots analyzing the impedance (Z) and phase angle (φ) responses, and modeled using an equivalent electrical circuit consisting of solution resistance (RΩ), double layer resistance (Rdl), polarization resistance (Rp), and associated capacitive elements, with impedance represented as complex values (real and imaginary components).

To assess tribocorrosion behavior, combined tribological and electrochemical testing was conducted in Ringer's lactate solution under reciprocating sliding motion. This setup simulated the mechanical and electrochemical conditions experienced by implants in vivo.

A custom-modified Nanovea T2000, manufactured by Nanovea, Irvine, CA, USA tribometer, integrated with an electrochemical cell, was used for the tribocorrsion tests, as shown in Figure 1. A three-electrode configuration was employed, consisting of an Ag/AgCl reference electrode, a platinum counter-electrode, and a working electrode comprising the coated sample (1 cm^2 exposed area), as depicted in Figure 1. The electrochemical data, including polarization curves and corrosion current densities, were recorded in real time using the Gamry Interface 1010E potentiostat.

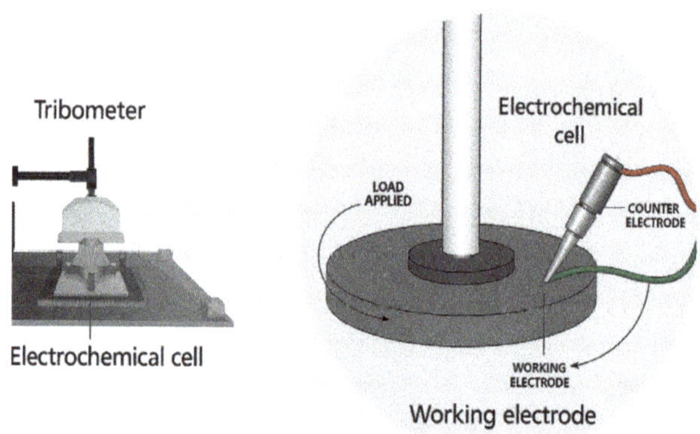

Figure 1. Experimental setup of the tribocorrosion test: reciprocating ball-on-flat tribometer coupled with an electrochemical cell.

Prior to testing, samples were immersed in Ringer's lactate solution for 120 min to stabilize the electrochemical interface. Subsequently, reciprocating sliding tests were conducted while simultaneously recording tribological and electrochemical responses. To ensure reproducibility, each test was repeated three times under identical experimental conditions.

3. Results and Discussion

3.1. Structural Characterization

Figure 2 presents four cross-sectional transmission electron microscopy (TEM) views of zirconium oxide coatings doped with silver (ZrO_2-Ag), deposited on silicon substrates. The coatings exhibit a dense and compact structure with no evident defects. A columnar microstructure is clearly visible, which can be attributed to bombardment by energetic particles (ions and neutrals) during the sputtering process. This bombardment enhances adatom mobility on the growing surface, thereby minimizing vacancy formation and promoting ordered growth.

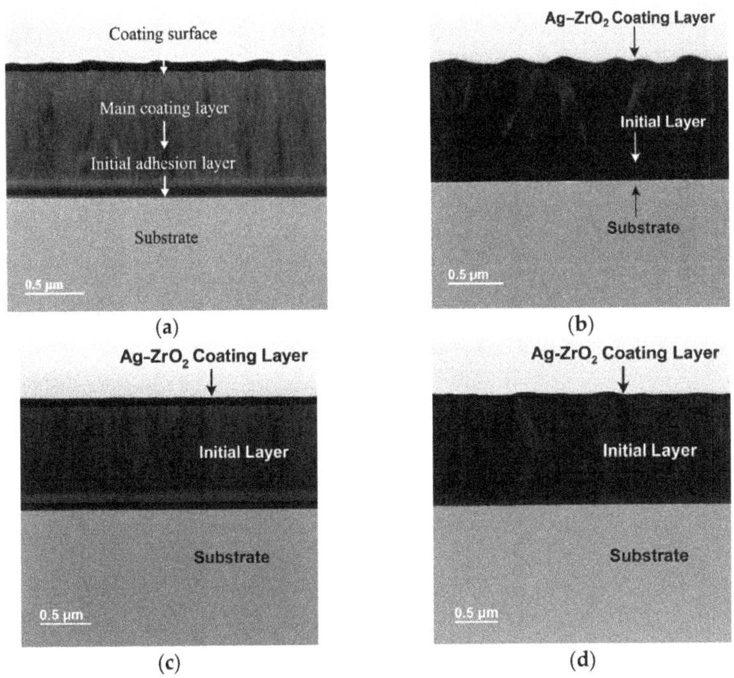

Figure 2. Cross-sectional TEM micrographs of ZrO_2-Ag coatings deposited with varying Ar/O_2 gas mixture ratios: (**a**) 12.5%, (**b**) 25%, (**c**) 37.5%, and (**d**) 50%.

The TEM micrographs display regions with varying contrast: brighter areas correspond to silver-enriched zones, while darker regions represent the ZrO_2 ceramic matrix. This contrast indicates a relatively homogeneous elemental distribution throughout the coating, in line with the established deposition parameters.

A distinct ZrO_2 adhesion-promoting interlayer, approximately 100 nm thick, is visible in all samples. This layer enhances adhesion between the coating and the substrate, as previously reported [20,28]. The images correspond to coatings deposited with different oxygen concentrations in the Ar/O_2 gas mixture: 12.5% (a), 25% (b), 37.5% (c), and 50% (d). The total coating thickness is approximately 1.2 µm, as confirmed via transmission electron microscopy measurements.

Figure 3 presents a morphological and compositional characterization of the ZrO_2-Ag coating synthesized under an Ar/O_2 atmosphere with 50% oxygen content. In part (a), the cross-sectional SEM image reveals a compact and homogeneous structure with a well-defined interface between the coating and the substrate. The observed microstructure suggests good densification and strong interfacial adhesion, attributes that are critical for enhanced mechanical performance and long-term protective behavior.

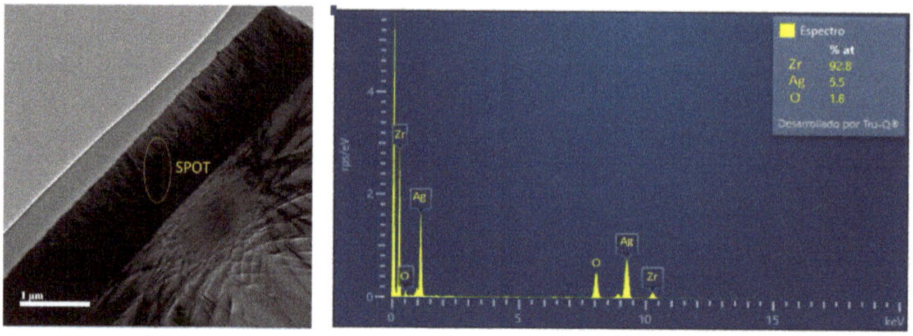

(a) (b)

Figure 3. (**a**) Cross-sectional SEM micrograph of the ZrO_2-Ag coating deposited under an Ar/O_2 atmosphere with 50% oxygen content. The area marked as "SPOT" indicates the region where the EDS analysis was performed. (**b**) Energy dispersive spectroscopy (EDS) acquired via scanning electron microscopy (SEM) for the sample deposited under an Ar/O_2 atmosphere with 50% oxygen content. The spectrum confirms the presence of Zr, O, and Ag in the coating.

A specific region within the coating, labeled as "SPOT", was selected for detailed elemental analysis. Part (b) shows the corresponding energy-dispersive spectroscopy (EDS) spectrum acquired from this targeted area. The spectrum displays prominent peaks associated with zirconium (Zr), oxygen (O), and silver (Ag), confirming the successful incorporation of these elements into the coating matrix. While EDS is a semi-quantitative technique, its primary utility in this context is to verify the presence of key elements rather than to provide precise compositional ratios. Nevertheless, the clear identification of Zr, O, and Ag supports the intended composition of the nanostructured coating and validates the deposition strategy.

Figure 4 presents the X-ray diffraction (XRD) pattern of the ZrO_2-Ag composite coatings, highlighting well-defined peaks associated with distinct crystalline phases. The Ar/O_2 gas ratio during deposition plays a critical role in influencing the crystallization behavior of zirconium oxide and the dispersion of metallic silver within the coating matrix [29,30]. Increasing the oxygen content up to 50% induces a phase transformation in the ZrO_2 coating, favoring the stabilization of the tetragonal phase. The 316LVM stainless steel substrate retained its characteristic austenitic crystalline structure, exhibiting face-centered cubic (FCC) crystallinity with diffraction peaks corresponding to the (111), (200), and (220) planes.

The XRD patterns in Figure 4 reveal well-defined peaks corresponding to both the monoclinic and tetragonal phases of ZrO_2, along with additional peaks corresponding to metallic silver (Ag), confirming the coexistence of oxide and metal phases within the composite. The crystal planes identified include (111), (101), (111), (112), (101), and (302), located at 2θ values of 27.95°, 30.32°, 32.09°, 37.93°, 44.98°, and 45.68°, respectively. These planes are characteristic of the Ag-ZrO_2 system, indicating successful integration between zirconia and silver. The presence of the (112) plane, in accordance with JCPDS card No. 89–7710 [31], is indicative of the tetragonal phase, while the monoclinic contributions align with JCPDS card No. 86–1451, suggesting that the deposition process promotes the formation of the tetragonal phase [32]. The (111) and (311) planes, with 2θ positions at 37.98° and 75.21°, correspond to the face-centered cubic (fcc) structure of metallic silver, aligning with JCPDS card No. 04–0783, which confirms that Ag particles are well crystallized within the composite matrix.

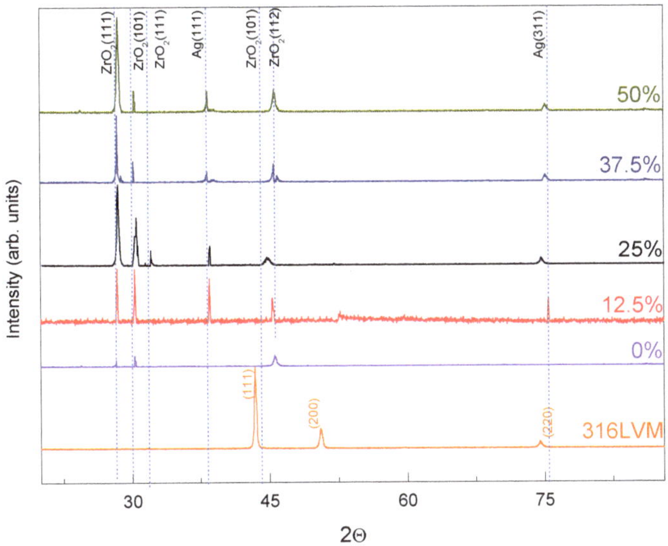

Figure 4. X-ray diffractogram of Ag-ZrO_2 coatings for different Ar/O_2 ratios.

An increase in the oxygen content during deposition enhances the crystallization of the tetragonal phase of ZrO_2, as evidenced by the increased intensity of the corresponding diffraction peaks. Variations in the relative intensities of Ag and ZrO_2 peaks are likely related to the distribution and dispersion of silver throughout the ceramic matrix, as well as the partial pressure of oxygen during film growth [33–36]. The simultaneous presence of Ag and ZrO_2 peaks without significant peak shifts indicates minimal chemical interaction between the phases, allowing each to retain its distinct crystalline structure.

3.2. Surface Properties and Topography

Figure 5 presents the surface topography of the ZrO_2-Ag coatings, as revealed by three-dimensional (3D) atomic force microscopy (AFM) imaging. The analysis indicates that the surface roughness (Ra) of the coatings varies as a function of the Ar/O_2 ratio employed during deposition. Notably, the sample deposited with Ar/O_2 −12.5% shows the highest average roughness (Figure 5a), with a Ra value of approximately 3.79 ± 0.12 nm, suggesting more pronounced granular growth and surface inhomogeneity [35]. As the oxygen concentration increases, a progressive decrease in surface roughness is observed: Ra $\approx 1.45 \pm 0.09$ nm for Ar/O_2 −25% (Figure 5b), Ra $\approx 1.10 \pm 0.06$ nm for Ar/O_2 37.5% (Figure 5c), and the lowest value of Ra $\approx 0.68 \pm 0.04$ nm at Ar/O_2 −50% (Figure 5d). This trend indicates that higher oxygen concentrations favor the formation of smaller grains and promote a more uniform and compact surface morphology, with reduced topographical irregularities [37,38]. Additionally, the observed reduction in roughness may also be attributed to the influence of silver on the nucleation and growth kinetics of the ZrO_2 matrix. The incorporation of Ag appears to facilitate the formation of a denser surface with fewer defects, likely by modifying the surface energy conditions and inhibiting abnormal grain growth during film formation.

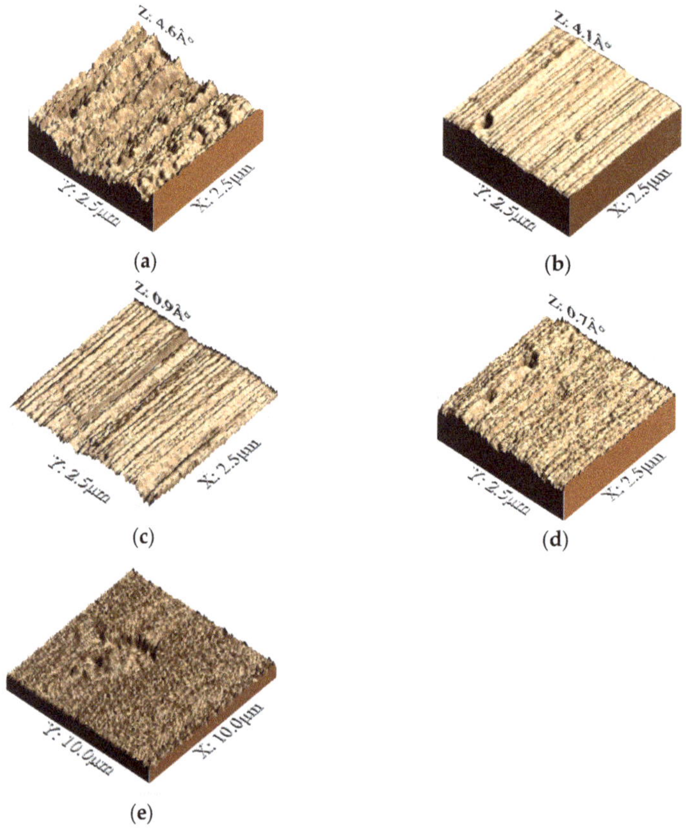

Figure 5. Three-dimensional topographic images of ZrO_2-Ag coatings deposited under different Ar/O_2 ratios, illustrating the influence of oxygen content on surface roughness and morphology. (**a**) 12.5%, (**b**) 25%, (**c**) 37.5%, and (**d**) 50%, (**e**) substrate 316LVM.

Figure 5 also includes three-dimensional AFM images of ZrO_2-Ag nanocomposite coatings deposited under varying Ar/O_2 ratios, providing additional qualitative insights into the effect of the oxygen content on surface morphology. Beyond the numerical roughness values, the images reveal a clear evolution in surface features with increasing oxygen content, from pronounced granular structures at 12.5% (Figure 5a) to smoother, more uniform surfaces at 50% (Figure 5d). These morphological changes are indicative of a shift in the grain growth mechanism and improved structural densification at higher oxygen concentrations. To further explore the influence of the coatings on substrate topography, AFM analysis was conducted over a larger scan area of 10×10 μm^2 (Figure 5e). The uncoated 316LVM substrate exhibits parallel surface features associated with the mechanical polishing process. In contrast, coated surfaces show distinct height distributions that correlate with the oxygen concentration: lower O_2 levels result in increased surface irregularities, whereas higher oxygen levels promote the development of a more compact and homogeneous morphology. These observations reinforce the conclusion that the oxygen concentration during deposition not only governs grain refinement, but also plays a critical role in controlling the surface texture and packing density of the coatings, factors that directly impact their tribological and electrochemical behavior.

3.3. Wear

Figure 6 illustrates the results of the wear tests, highlighting a notable improvement in the tribological performance of the Ag-doped ZrO_2 coatings as the oxygen content in the Ar/O_2 deposition atmosphere increases. A higher oxygen ratio (12.5%, 25%, 37.5%, and 50%) correlates with a progressive reduction in both the average coefficient of friction (COF) and in the wear volume, indicating enhanced wear resistance, particularly for coatings deposited at a 50% Ar/O_2 ratio. In this condition, the Ar/O_2 50% ratio coating exhibits minimal wear and reduced plastic deformation, consistent with a transition from adhesive to mild abrasive wear mechanisms, as evidenced by shallower grooves and smoother wear track.

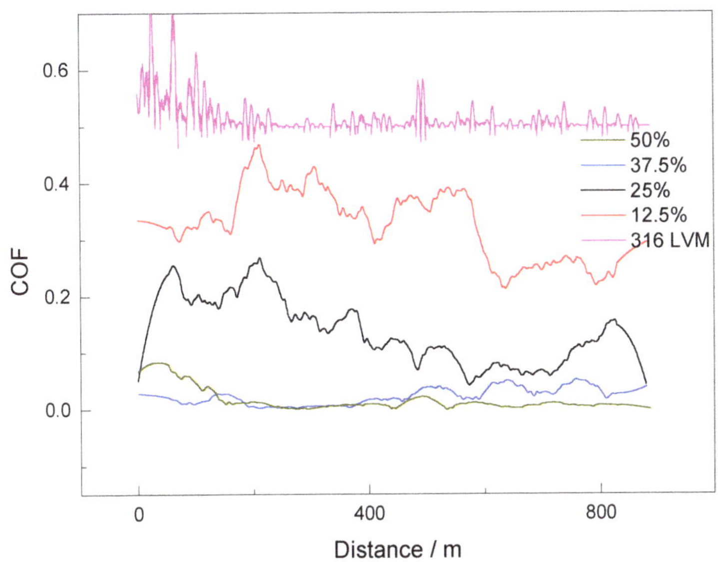

Figure 6. Coefficient of friction as a function of sliding distance for Ag-doped ZrO_2 coatings deposited with different Ar/O_2 ratios. The curves correspond to 50% (olive), 37.5% (blue), 25% (black), 12.5% (red), and uncoated 316LVM substrate (violet).

The coating deposited under an Ar/O_2 atmosphere containing 50% oxygen exhibited the lowest coefficient of friction (0.0043) and the smallest wear volume (0.086 mm^3). This behavior is attributed to improved phase consolidation, which enhances the coating's abrasion resistance and facilitates the formation of a dense, adherent layer with potential lubricating properties [39]. Conversely, coatings grown under lower oxygen concentrations, particularly 12.5%, show poor structural refinement, leading to inhomogeneous particle distribution, pronounced plastic deformation, and deeper grooves, confirming more severe wear, as corroborated by AFM analyses.

Coatings obtained at intermediate O_2 ratios (25% and 37.5%) display transitional behavior, with moderate wear resistance improvements. The sample prepared at 37.5% O_2 achieved a COF of 0.032, significantly higher than that of the 50% O_2 coating, despite exhibiting similar roughness values. This suggests that the friction behavior is influenced more by mechanical properties than by surface topography alone. These intermediate coatings indicate that increasing the oxygen content contributes to a more efficient deposition process and improved microstructure, enhancing wear resistance relative to the 12.5% O_2 sample.

During sliding, the adaptation of the softer counter body (pin) to the surface morphology of the ZrO_2-Ag coatings prepared at 50% Ar/O_2 results in a gradual decrease in COF, reaching a steady state after approximately 150 m of sliding [40]. This stabilization is attributed to the interplay between adhesive and interfacial mechanisms, with the final COF primarily governed by the elastoplastic response and surface roughness of the coating. Coatings deposited at lower O_2 ratios fail to reach this steady state, exhibiting fluctuating COF profiles due to unstable contact conditions.

To quantify wear resistance, the Archard equation was employed [41], relating the volume of the material removed to the applied normal load and sliding distance. The wear volumes, calculated from profilometric measurements of the wear tracks, confirmed the inverse relationship between the oxygen content and wear rate. For instance, coatings deposited with 12.5% and 25% O_2 showed wear volumes of 1.072 mm^3 and 0.086 mm^3, respectively, while those deposited at 37.5% and 50% O_2 exhibited significantly reduced volumes of 0.248 mm^3 and 0.086 mm^3, respectively. These results underscore that increasing the oxygen concentration during deposition leads to dense and more uniform coatings, thereby reducing material loss under sliding conditions [42]. This suggests that a lower oxygen concentration in the deposition atmosphere hinders the formation of a stable and homogeneous ZrO_2 layer, negatively impacting the wear resistance of the coating. Under these conditions, ZrO_2 particles tend to be less uniformly distributed, leading to higher wear rates during tribological tests, with friction coefficients reaching 0.16 for 25% and 0.38 for 12.5% ratios. These coatings exhibit a more aggressive wear mechanism, favoring plastic deformation and deep groove formation.

Coatings with 25% and Ar/O_2 37.5% gas mixtures show a transition in their tribological properties, with an evident but less pronounced improvement in wear resistance compared to the 50% coating. In these cases, a higher oxygen content in the atmosphere facilitates a more efficient deposition process and better microstructure, contributing to a more homogeneous ZrO_2 particle distribution. This allows the coating to maintain more efficient and durable wear behavior compared to Ar/O_2 12.5% coatings.

The tribological profiles also reveal that COF fluctuations are strongly linked to the dominant wear mechanisms. For example, the 316LVM substrate exhibits significant COF variability due to the accumulation of wear debris and unstable contact. Repeatability was confirmed by conducting three wear tests per sample, all yielding consistent COF trends in both amplitude and curve morphology, indicating reliable tribological performance under the experimental conditions.

Finally, considering the relationship between COF, applied load, and sliding distance, it can be inferred that as long as the applied load remains below the critical threshold for surface failure, the frictional behavior stabilizes and the COF decreases due to improved contact conditions and coating integrity.

3.4. Electrochemical Evaluation

Figure 7 shows the Bode plots of the ZrO_2-Ag coatings, revealing the electrochemical response of each sample across a frequency range from 100 kHz to 0.01 Hz. An increase in the magnitude of the impedance modulus ($|Z|$) is observed for coatings deposited with a higher oxygen content, particularly in the low-frequency region. This behavior indicates enhanced barrier properties and improved resistance to electrolyte penetration.

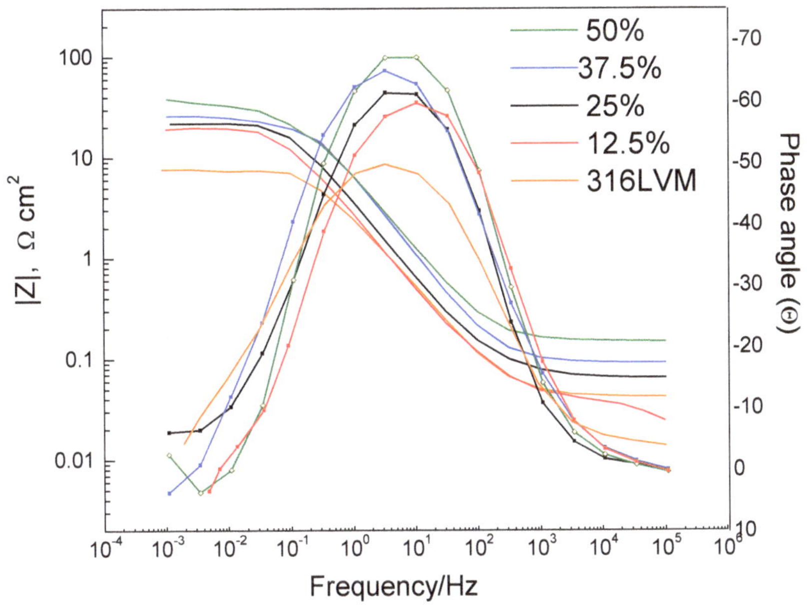

Figure 7. Bode plots showing impedance magnitude and phase angle as a function of frequency for ZrO_2-Ag coatings deposited with different Ar/O_2 ratios. Curves correspond to 50% (green), 37.5% (blue), 25% (black), 12.5% (red), and uncoated 316LVM substrate (orange).

The plots display characteristics consistent with the presence of two distinct time constants, suggesting the contribution of two interfacial processes: one associated with the outer coating–electrolyte interface and the other with the inner coating–substrate interface. The increased phase angle and broader plateau in the mid-to-low frequency region for the coatings deposited at 37.5% and 50% O_2 are indicative of superior electrochemical stability and capacitive behavior.

These results confirm that oxygen concentrations during deposition promote the formation of more compact and homogeneous coatings, which effectively hinder charge transfer and ionic diffusion. Consequently, coatings synthesized under higher Ar/O_2 ratios demonstrate optimized protective performance against electrochemical degradation in simulated physiological environments [43].

Analysis of the Bode plots confirms a progressive decrease in total impedance as the Ar/O_2 ratio in the deposition atmosphere is reduced (from 50% to 12.5%). This trend is attributed to the less effective formation of a continuous and compact oxide layer, which alters the coating's electrochemical behavior and compromises its protective capabilities [39]. The impedance response across the frequency spectrum enables the detailed characterization of the system's charge transfer resistance and interfacial capacitance, two key parameters for evaluating coating stability under simulated physiological conditions.

From an electrochemical impedance spectroscopy (EIS) standpoint, high- and medium-frequency regions reflect processes occurring at the coating–electrolyte interface, whereas medium to low frequencies are dominated by charge transfer kinetics and ionic diffusion mechanisms. At the lowest frequencies, the impedance response is primarily influenced by interactions between the coating and the underlying substrate. The presence of a partial semicircle in the Nyquist representation indicates surface-controlled conduction, whose diameter diminishes as the oxygen content decreases. This reduction suggests that denser and more uniform films, typically obtained at higher oxygen ratios, enhance the overall system impedance and thereby improve corrosion resistance.

Coatings synthesized under higher Ar/O_2 ratios display markedly higher impedance magnitudes, in agreement with previous tribomechanical findings, where these systems also demonstrated superior wear resistance. This correlation implies that controlled oxygen incorporation during deposition promotes the formation of dense, adherent, and electrochemically stable coatings, contributing to improved durability under aggressive environmental conditions.

Additionally, variations in time constants derived from the EIS spectra suggest that the electrochemical behavior of the coatings evolves with prolonged exposure to the corrosive medium. In the ZrO_2-Ag system, shifts in interfacial capacitance are observed, potentially related to the passivation of exposed substrate regions in areas where the coating may contain minor discontinuities [44]. This self-passivation process favors the formation of a protective oxide layer, thereby enhancing corrosion resistance and improving coating–substrate adhesion.

The slight displacement of the time constants over time also indicates modifications in the dielectric response due to both the coating and the substrate, likely associated with progressive passivation effects. As shown in the equivalent circuit model (Figure 8), two distinct time constants are observed during the first hour of immersion, with the second remaining stable thereafter, suggesting that the system reaches electrochemical equilibrium. Furthermore, the low root mean square (RMS) roughness values observed through AFM support this interpretation, indicating that a homogeneous surface morphology contributes to improving resistance to electrochemical degradation.

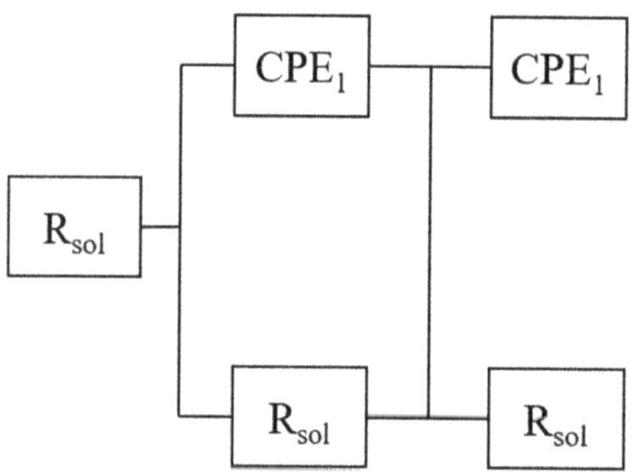

Figure 8. Equivalent circuit used to model the electrochemical response of the coatings, considering electrolyte resistance (Rsol), coating resistance (Rpo), charge transfer resistance (Rct), and constant phase elements (CPE_1 and CPE_2).

Two sets of constant phase elements (CPEs) in parallel were incorporated into the equivalent circuit model to simulate the dielectric and electrochemical behavior of the ZrO_2-Ag coatings. The first pair (CPE_1 and Rpo) represents the barrier properties of the coating. Here, Rpo accounts for the pore resistance, indicating the coating's ability to resist electrolyte penetration through structural defects such as pores or microcracks. CPE_1 describes the non-ideal capacitive behavior of the coating–electrolyte interface.

The second pair (CPE_2 and Rct) simulates the electrochemical processes occurring at the metal–coating interface, primarily through localized defects. The Rct (charge transfer resistance) reflects the kinetics of the corrosion reaction, while CPE_2 captures the dielectric properties of the double layer formed at these active sites.

In the equivalent circuit model, Rsol represents the solution resistance between the working and reference electrodes. The use of constant phase elements allows for the more accurate modeling of non-ideal capacitive behavior, typically observed in real electrochemical systems with surface roughness, inhomogeneities, or interfacial porosity. This modeling approach enables a refined interpretation of the charge transfer and mass transport phenomena governing coating performance under corrosive conditions.

The fitted parameter values for the ZrO_2-Ag coatings are consistent with the trends observed in the Bode plots. Notably, both the Rpo and Rct increase with higher Ar/O_2 ratios (12.5%, 25%, 37.5%, and 50%), suggesting the formation of a more compact, stable, and protective passive oxide layer. An increase in the Rct reflects a slower charge transfer process and enhanced corrosion resistance, while a higher Rpo value indicates reduced electrolyte access to the substrate through coating defects [45].

Table 1 summarizes the electrical parameters obtained from fitting the EIS data. The solution resistance (Rsol) shows minor variation across the coated samples, ranging from 45 to 64 $\Omega\cdot cm^2$. In contrast, the uncoated 316LVM substrate exhibits a significantly lower Rsol value (12.05 $\Omega\cdot cm^2$), likely due to its direct exposure to the electrolyte, which facilitates ionic exchange. The double-layer capacitance values (C_1 and C_2), along with their associated dispersion factors (α_1 and α_2), provide insights into the coatings' dielectric properties and surface uniformity. Notably, C_2 decreases steadily as the oxygen content increases, from 610.24 $\mu F\cdot cm^{-2}$ (12.5% Ar/O_2) to 287.39 $\mu F\cdot cm^{-2}$ (50% Ar/O_2), suggesting the formation of a denser and less porous passive film.

Table 1. Parameters used in the simulation of impedance data.

Ar/O_2	R_s Ω cm^2	C_1 μF cm^{-2}	$\alpha 1$	R_1 $k\Omega$ cm^2	C_2 μF cm^{-2}	$\alpha 2$	R_2 $k\Omega$ cm^2
12.5%	55.06 ± 2.75	45.28 ± 2.26	0.88 ± 0.02	3.56 ± 0.18	610.24 ± 30.5	0.84 ± 0.02	8.14 ± 0.41
25%	64.03 ± 3.20	39.41 ± 1.97	0.89 ± 0.03	4.62 ± 0.23	504.36 ± 25.2	0.82 ± 0.03	9.47 ± 0.47
37.5%	53.04 ± 2.65	35.37 ± 1.77	0.87 ± 0.02	5.8 ± 0.29	344.45 ± 17.22	0.91 ± 0.03	14.52 ± 0.73
50%	45.02 ± 2.25	32.41 ± 1.62	0.95 ± 0.02	6.8 ± 0.34	287.39 ± 14.3	0.96 ± 0.02	38.83 ± 1.94
316LVM	12.05 ± 0.60	14.57 ± 0.73	0.84 ± 0.03	3.14 ± 0.16	205.32 ± 10.2	0.89 ± 0.02	5.98 ± 0.30

The most critical indicators of corrosion resistance, R_1 and R_2, exhibit a clear increasing trend with a higher oxygen content. Specifically, R_2 rises from 8.14 $k\Omega\cdot cm^2$ (12.5%) to 38.83 $k\Omega\cdot cm^2$ (50%), reflecting enhanced charge transfer resistance and superior barrier performance. This trend is inversely correlated with the C_2 values, reinforcing the conclusion that increased oxygen during deposition promotes the formation of a compact and electrochemically stable oxide layer. The highest R_2 and α_2 values observed in the 50% Ar/O_2 coating further confirm the optimized microstructure and improved homogeneity of this system.

3.5. Triboelectrochemical Wear on Corrosion Resistance

Figure 9 presents the electrochemical behavior of the ZrO_2-Ag coatings under tribocorrosive conditions, represented through Tafel plots. These curves allow for the evaluation of the synergistic effects between wear and corrosion and illustrate how the electrochemical response varies with different Ar/O_2 gas ratios during deposition. The sample deposited under the highest Ar/O_2 50% ratio exhibits the most favorable electrochemical performance, characterized by the lowest corrosion current density (Icorr) and the highest polarization resistance (Rp). This indicates that the coating provides superior protection against corrosion, even under mechanical stress [46,47]. This enhanced behavior is attributed to the formation of a dense and chemically stable ZrO_2 passive layer, which serves as an effective barrier to aggressive ionic species and simultaneously improves wear resistance. The high oxygen content during deposition likely promotes better crystallinity and homogeneity of the ceramic matrix, thereby enhancing its electrochemical stability.

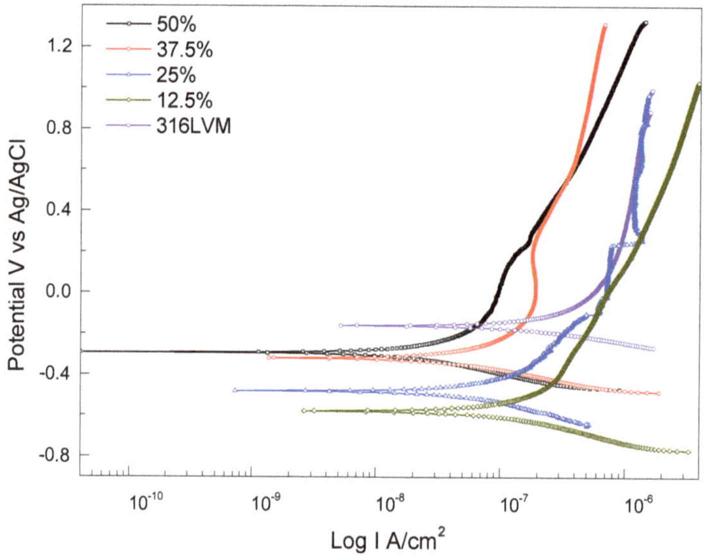

Figure 9. Tafel curves obtained from electrochemical tests, illustrating the synergy between wear and corrosion in ZrO_2-Ag coatings with different Ar/O_2 ratios. Curves are distinguished as follows: 50% (□, solid black), 37.5% (○, solid red), 25% (△, solid blue), 12.5% (◇, solid olive), and 316LVM (○, solid violet).

In contrast, the coating deposited with an Ar/O_2 12.5% ratio demonstrates the poorest electrochemical performance, with significantly higher corrosion current density values and reduced polarization resistance, as summarized in Table 2. This inferior behavior is indicative of a less protective and possibly discontinuous passive layer, which is more susceptible to breakdown under tribological loading [48]. The absence of a compact and robust passive layer increases its vulnerability to localized corrosion, ultimately accelerating material degradation.

Table 2. Electrochemical parameters obtained from Tafel analysis for ZrO_2-Ag coatings deposited under different Ar/O_2 ratios, highlighting the influence of deposition conditions on corrosion resistance.

Ar/O_2	12.5%	25%	37.5%	50%	316LVM
Icorr (A/cm^2)	1.92×10^{-7}	9.99×10^{-8}	6.77×10^{-8}	3.32×10^{-8}	6.02×10^{-7}
Ecorr (V)	−0.57	−0.47	−0.32	−0.28	−0.13
Vcorr mm/y	0.00209	0.00109	0.00074	0.00036	0.0067
% IE	68.11	83.41	88.75	94.49	-

The Tafel analysis clearly demonstrates that the quality of the coating significantly affects the wear–corrosion synergy [48]. Coatings with more robust and stable passive layers not only minimize corrosion, but also reduce the coefficient of friction and wear volume. Therefore, coatings produced under a higher oxygen content exhibit superior triboelectrochemical performance, both in terms of mechanical durability and corrosion resistance, compared to those deposited at lower oxygen levels.

Additionally, the incorporation of silver into the ZrO_2 matrix appears to contribute to enhanced corrosion resistance through a sacrificial anodic mechanism. In chloride-rich environments, such as Ringer's lactate solution, silver undergoes preferential oxidation, offering cathodic protection to the underlying stainless steel substrate and supporting the stability of the passive layer on the coating surface. This anodic contribution explains the lower corrosion current densities and higher charge transfer resistance observed in samples with an increased silver content and oxygen ratio. Although silver may also act as a solid lubricant, its electrochemical function as a sacrificial component is considered the primary mechanism contributing to the improved tribocorrosive performance.

The inhibition efficiency (IE%) was also calculated based on the corrosion current density, using the uncoated sample (316LVM) as the reference. The equation used was as follows:

$$IE\% = [1 - (Icorr\_coated/Icorr\_uncoated)] \times 100$$

The results, presented in Table 2, demonstrate a significant improvement in corrosion protection for all coated samples. The trend indicates that increasing the oxygen content in the deposition atmosphere enhances the inhibition efficiency, demonstrating both the effectiveness of silver doping and the beneficial influence of oxygen on the coating performance.

These results are consistent with the observed corrosion rate and corrosion potential values, confirming that a higher oxygen content favors the formation of more stable, dense, and protective coatings.

Therefore, the results suggest that optimizing the Ar/O_2 gas ratio in the deposition atmosphere is critical for enhancing the corrosion and wear resistance of ZrO_2-Ag coatings, with the added benefit of protecting the substrate from triboelectrochemical degradation.

The Ag-doped ZrO_2 coatings deposited under varying Ar/O_2 gas ratios demonstrated clear passivation behavior, characterized by low corrosion current densities. Notably, coatings deposited at a higher Ar/O_2 ratio of 50% (Table 2) exhibited the lowest current densities, indicating the formation of a smoother and more compact surface that restricts ionic penetration and minimizes corrosion through oxidation reactions. In contrast, coatings with increased surface roughness, often resulting from a lower oxygen content, can facilitate crack initiation and propagation, thereby increasing susceptibility to mechanical failure and localized corrosion. The coatings produced at higher Ar/O_2 ratios also displayed more positive corrosion potentials and lower corrosion current densities, further supporting their enhanced protective capacity.

Moreover, the incorporation of silver into the coating matrix appears to influence the electrochemical behavior. A general trend of decreasing current density with increasing silver content was observed, potentially due to redox reactions involving metallic silver at the coating surface. The Tafel curves for Ag-containing coatings showed fluctuations in the current density values, indicating the presence of multiple concurrent electrochemical processes. These likely involve chloride or lactate anions interacting with both the exposed metallic substrate and silver particles at the surface.

A comprehensive assessment of triboelectrochemical behavior, evaluating the simultaneous action of mechanical wear and electrochemical corrosion, provides deeper insight into the coating performance under realistic service conditions. During tribocorrosion tests conducted in Ringer's lactate solution, ZrO_2-Ag coatings outperformed uncoated 316LVM stainless steel substrates. This superior behavior is attributed to the synergistic properties of the ZrO_2, known for its excellent corrosion resistance and hardness, and silver, which acts as both a solid lubricant and a sacrificial anode. Electrochemical measurements using a three-electrode system confirmed this enhancement, with coated samples exhibiting significantly lower corrosion current densities and higher polarization resistances under sliding wear conditions.

The results of this study highlight a significant synergistic effect between the Ar/O_2 deposition ratio and the resulting triboelectrochemical performance of ZrO_2-Ag coatings. Coatings fabricated at a high Ar/O_2 50% ratio demonstrated the most favorable performance, with a remarkably low coefficient of friction (0.0043), minimal wear volume (0.086 mm^3), and a corrosion current density of 3.32×10^{-8} A/cm^2. These improvements are attributed to a highly consolidated ZrO_2 phase structure, which enhances mechanical integrity, along with a homogeneous distribution of silver that contributes both lubricating and antioxidative functionalities. Conversely, coatings produced with lower Ar/O_2 12.5% ratios showed inferior performance, dominated by adhesive wear mechanisms, higher friction coefficients, and greater corrosion susceptibility due to their inhomogeneous microstructure.

The observed interplay between mechanical wear and electrochemical stability highlights the critical importance of controlling deposition parameters to engineer coatings with tailored multifunctional properties. This is particularly relevant for demanding applications such as biomedical implants and marine components, where materials are routinely exposed to simultaneous mechanical and corrosive stressors. Future studies should explore the long-term performance of ZrO_2-Ag coatings under cyclic tribocorrosion conditions to further validate their durability and clinical viability.

When compared to other protective coatings such as TiCN, TiSiCN, TiCrSiCN, TiAlSiCN, and ReN, the ZrO_2-Ag system reveals superior tribocorrosive performance. Conventional TiCN and TiSiCN coatings, although widely employed, are prone to accelerated wear and brittle fracture under dynamic loading due to structural defects and limited plasticity [49]. Their Cr- and Al-doped counterparts (TiCrSiCN and TiAlSiCN) offer moderate improvements in mechanical performance, but remain susceptible to microcracks and delamination caused by microstructural inhomogeneities. In contrast, the ReN coating have demonstrated outstanding corrosion resistance, particularly as the N_2 partial pressure during deposition increases. This behavior is associated with denser film formation and more efficient nitrogen incorporation, resulting in corrosion current density (Icorr) reductions of up to 98.94% [50]. However, while ReN coatings perform well electrochemically, their tribological properties under combined mechanical and corrosive conditions require further validation.

The proposed ZrO_2-Ag coating, on the other hand, offers an ideal combination of high corrosion resistance and mechanical durability. Its wear mechanism is governed primarily by controlled plastic deformation, which significantly reduces volume loss. Additionally, the in situ formation of a passive ZrO_2 layer, reinforced by the uniformly distributed silver phase, provides sustained protection even in chloride-rich environments. This dual functionality makes the ZrO_2-Ag coating a promising and robust alternative to conventional protective systems for use in biomedical and other aggressive application settings.

The influence of the Ar/O_2 gas mixture ratio on the performance of Ag-ZrO_2 coatings is a key parameter that directly impacts the microstructural, tribological, and electrochemical characteristics of the deposited films. Although our results show a consistent improvement in wear resistance and inhibition efficiency with increasing oxygen content, the interpretation of these effects requires further context. In this regard, the observed enhancement in corrosion protection for coatings deposited with higher oxygen concentrations (e.g., 50% O_2) can be attributed to the formation of a more homogeneous and fully oxidized zirconia matrix, as also reported by Zhang et al. (2007) [51], who found that high oxygen partial pressures promoted the stabilization of the tetragonal phase, leading to denser and more protective oxide films.

Additionally, previous studies on Ag-doped ZrO_2 coatings deposited via reactive sputtering (e.g., Ou et al., 2015) [52] support the notion that increased oxygen flow during deposition favors the better dispersion of silver nanoparticles and suppresses porosity, thereby enhancing the electrochemical barrier properties of the coating. Our results are in agreement with these findings, as evidenced by the increasing charge transfer resistance and decreasing corrosion current densities with the higher O_2 content. Furthermore, the improved crystallinity of Ag, confirmed via XRD patterns showing sharper (111) and (311) peaks at higher oxygen ratios, aligns with the conclusions of Biju et al. (2025) [53], who demonstrated that oxygen-rich environments stabilize metallic Ag domains within oxide matrices.

However, it is also important to note that excessively high oxygen flow rates can, in some cases, lead to over-oxidation or poor adhesion due to reduced energy transfer during sputtering. In our study, no such detrimental effects were observed up to 50% O_2, suggesting that the selected range of gas mixtures remained within an optimal processing window. This finding underscores the importance of the precise control of the oxygen content during deposition to balance phase formation, coating integrity, and functional performance.

4. Conclusions

The Ar/O_2 ratio during deposition significantly influences the microstructure, topography, and tribological performance of Ag-ZrO_2 coatings. A higher oxygen content promotes the better crystallization of zirconium oxide, leading to reduced surface roughness and enhanced wear resistance. These characteristics suggest potential applications in systems requiring high durability and low friction coefficients.

X-ray diffraction analysis confirmed the presence of monoclinic and tetragonal phases of ZrO_2, as well as crystallized metallic silver. The tetragonal phase was more prevalent at higher Ar/O_2 ratios, indicating that the deposition process modulates the crystalline structure of the coating.

Surface roughness measurements indicated that coatings deposited with a higher Ar/O_2 ratio exhibited lower Ra values, resulting in a more compact and homogeneous morphology. This suggests that the nucleation and grain growth processes are strongly affected by the deposition atmosphere.

Tribological tests demonstrated that coatings deposited with a higher Ar/O_2 ratio exhibited lower friction coefficients and reduced volume loss, indicating improved wear resistance. The best tribological performance was observed in coatings with 50% Ar/O_2, which showed the lowest volume loss and friction coefficient.

Electrochemical tests revealed that coatings deposited with higher Ar/O_2 ratios exhibited lower corrosion current densities and higher polarization resistance. This improvement is attributed to the formation of a stable and dense passive ZrO_2 layer, which enhances the coating's resistance to corrosion.

The optimized Ag-ZrO_2 coatings developed in this study demonstrate potential for biomedical applications, particularly in implants where enhanced durability and corrosion resistance are critical.

Future research should focus on long-term in vitro and in vivo studies to evaluate the biocompatibility and stability of these coatings in physiological environments.

Author Contributions: Conceptualization and methodology, G.O.-H. and W.A.; validation, W.A.; formal analysis, G.O.-H.; investigation, W.A. All authors have read and agreed to the published version of the manuscript.

Funding: This research received no external funding.

Institutional Review Board Statement: Not applicable.

Informed Consent Statement: Not applicable.

Data Availability Statement: The initial characterization data are reported in the following repository: IOP Publishing dataset. https://iopscience.iop.org/article/10.1088/1742-6596/687/1/012031.

Acknowledgments: W. Aperador acknowledges the support of Universidad Militar Nueva Granada. G. Orozco acknowledges the support of Universidad ECCI.

Conflicts of Interest: The authors declare no conflicts of interest.

References

1. Maver, U.; Xhanari, K.; Žižek, M.; Gradišnik, L.; Repnik, K.; Potočnik, U.; Finšgar, M. Carboxymethyl cellulose/diclofenac bioactive coatings on AISI 316LVM for controlled drug delivery, and improved osteogenic potential. *Carbohydr. Polym.* **2020**, *230*, 115612. [CrossRef] [PubMed]
2. Pachla, W.; Skiba, J.; Kulczyk, M.; Przybysz, S.; Przybysz, M.; Wróblewska, M.; Diduszko, R.; Stępniak, R.; Bajorek, J.; Radomski, M.; et al. Nanostructurization of 316L type austenitic stainless steels by hydrostatic extrusion. *Mater. Sci. Eng. A* **2014**, *615*, 116–127. [CrossRef]
3. Carreon, H.; Barriuso, S.; Barrera, G.; González-Carrasco, J.L.; Caballero, F.G. Assessment of blasting induced effects on medical 316 LVM stainless steel by contacting and non-contacting thermoelectric power techniques. *Surf. Coat. Technol.* **2012**, *206*, 2941–2946. [CrossRef]
4. López, R.; Menéndez, M.; Fernández, C.; Chmiela, A.; Bernardo-Sánchez, A. The Influence of Carbon Coatings on the Functional Properties of X39Cr13 and 316LVM Steels Intended for Biomedical Applications. *Metals* **2019**, *9*, 815. [CrossRef]
5. Lo, K.H.; Shek, C.H.; Lai, J.K.L. Recent developments in stainless steels. *Mater. Sci. Eng. R Rep.* **2009**, *65*, 39–104. [CrossRef]
6. Hamedi, H.; Isfahani, T. Wear and corrosion properties of mechanically coated 316 stainless Steel-TiC nanocomposites. *Results Eng.* **2024**, *24*, 102966. [CrossRef]
7. Freitas Filho, A.; Silva, G.C.; Rodrigues, S.C.S.; Santos, A.J. Evaluation of the effect of surface modification of Ti64 and 316 L by addition of calcium phosphate through electrical discharge machining process. *Tribol. Int.* **2023**, *180*, 108245. [CrossRef]
8. Thambapillary, S.; Dimitriou, R.; Makridis, K.G.; Fragkakis, E.M.; Bobak, P.; Giannoudis, P.V. Implant longevity, complications and functional outcome following proximal femoral arthroplasty for musculoskeletal tumors: A systematic review. *J. Arthroplast.* **2013**, *28*, 1381–1385. [CrossRef] [PubMed]
9. Nasiri-Tabrizi, B.; Basirun, W.J.; Walvekar, R.; Yeong, C.H.; Phang, S.W. Exploring the potential of intermetallic alloys as implantable biomaterials: A comprehensive review. *Biomater. Adv.* **2024**, *161*, 213854. [CrossRef] [PubMed]
10. Zhang, E.; Zhao, X.; Hu, J.; Wang, R.; Fu, S.; Qin, G. Antibacterial metals and alloys for potential biomedical implants. *Bioact. Mater.* **2021**, *6*, 2569–2612. [CrossRef] [PubMed]
11. Elsa, G.; Hanan, A.; Walvekar, R.; Numan, A.; Khalid, M. Zirconium–Based MXenes: Synthesis, Properties, Applications, and Prospects. *Coord. Chem. Rev.* **2025**, *526*, 216355. [CrossRef]
12. Harun-Ur-Rashid, M.; Foyez, T.; Krishna, S.B.N.; Poda, S.; Imran, A.B. Recent Advances of Silver Nanoparticle-Based Polymer Nanocomposites for Biomedical Applications. *RSC Adv.* **2025**, *15*, 8480–8505. [CrossRef] [PubMed]
13. Yang, C.; Wang, C.; Shen, Z.; Zhou, L.; Sheng, L.; Xu, D.; Zheng, Y.; Chu, P.K.; Xiao, S.; Ying, T.; et al. Simultaneous Improvement of Wear and Corrosion Resistance of Microarc Oxidation Coatings on ZK61 Mg Alloy by Doping with ZrO_2 Nanoparticles. *J. Mater. Sci. Technol.* **2025**, *224*, 312–327. [CrossRef]
14. Li, Z.-Y.; Yu, J.-G.; Ren, Q.-Y.; Zheng, M.-Y.; Cai, Z.-B., Jiao, Y.-J. Study on the Fretting Corrosion Behavior of Zirconium Alloy in Simulated Primary Coolant Condition. *Wear* **2025**, *570*, 206057. [CrossRef]
15. Niu, D.; Zhang, C.; Sui, X.; Lu, X.; Zhang, X.; Wang, C.; Hao, J.; Shi, Z. Microstructure, mechanical properties and tribo-corrosion mechanism of (CrNbTiAlVMo) coated 316L stainless steel in 3.5 wt% NaCl solution. *Tribol. Int.* **2022**, *173*, 107638. [CrossRef]
16. Davoodi, F.; Atapour, M.; Ashrafizadeh, F.; Rikhtehgaran, R. Dry sliding wear characteristics of NiP/TiN duplex coated aluminium alloy and wear analysis using response surface method. *J. Mater. Eng. Perform.* **2022**, *31*, 6360–6372. [CrossRef]
17. Fook, P.; Berger, D.; Riemer, O.; Karpuschewski, B. Structuring of Bioceramics by Micro-Grinding for Dental Implant Applications. *Micromachines* **2019**, *10*, 312. [CrossRef] [PubMed]
18. Li, X.; Liang, S.; Li, J.; Tang, W.; Yu, M.; Ahmed, M.H.; Liang, S.; Zhang, F.; Inokoshi, M.; Yao, C.; et al. Influence of surface treatments on highly translucent zirconia: Mechanical, optical properties and bonding performance. *J. Dent.* **2025**, *154*, 105580. [CrossRef] [PubMed]
19. Zhang, F.; Van Meerbeek, B.; Vleugels, J. Importance of tetragonal phase in high-translucent partially stabilized zirconia for dental restorations. *Dent. Mater.* **2020**, *36*, 491–500. [CrossRef] [PubMed]
20. Li, X.; He, J.; Zhang, W.; You, L.; Li, J. Ag nanoparticles interlayered Fe_3O_4/Ag/m(TiO_2-ZrO_2) magnetic photocatalysts with enhanced stability and photocatalytic performance for Cr(VI) reduction. *Appl. Surf. Sci.* **2023**, *607*, 155076. [CrossRef]
21. Shao, Y.; Jiang, Y.; Wang, Y.; Dong, Q.; Wang, C.; Wang, Y.; Feng, X.; Chu, C.; Bai, J. Electrodepositing Ag on anodized stainless steel for enhanced antibacterial properties and corrosion resistance. *J. Funct. Biomater.* **2025**, *16*, 19. [CrossRef] [PubMed]

22. Aissi, M.; Tayyaba, Q.; Er-Ramly, A.; Hermawan, H.; Merzouk, N. Improving the clinical performance of dental implants through advanced surface treatments: The case of Ti and ZrO_2 coatings. *Metals* **2025**, *15*, 320. [CrossRef]
23. Bai, L.; Yi, G.; Wan, S.; Wang, W.; Sun, H. Comparison of tribological performances of plasma sprayed YSZ, YSZ/Ag, YSZ/MoO_3 and YSZ/Ag/MoO_3 coatings from 25 to 800 °C. *Wear* **2023**, *526–527*, 204944. [CrossRef]
24. González-Hernández, A.; Aperador, W.; Flores, M.; Onofre-Bustamante, E.; Bermea, J.E.; Bautista-García, R.; Gamboa-Soto, F. Influence of Deposition Parameters on Structural and Electrochemical Properties of Ti/Ti_2N Films Deposited by RF-Magnetron Sputtering. *Metals* **2022**, *12*, 1237. [CrossRef]
25. Guzmán, P.; Yate, L.; Sandoval, M.; Caballero, J.; Aperador, W. Characterization of the Micro-Abrasive Wear in Coatings of TaC-HfC/Au for Biomedical Implants. *Materials* **2017**, *10*, 842. [CrossRef] [PubMed]
26. Fang, M.; Liu, B.; Wang, Y.; Yang, Z.; Wang, Y. Interfacial microstructure and mechanical properties of joints between ZrO_2 ceramic and 316 stainless steel brazed using Ag–Cu–In–Ti filler. *Vacuum* **2023**, *218*, 112668. [CrossRef]
27. Irshad, M.; Ibrahim, M.M.; Siddique, S.; Younas, U.; Mersal, G.A.M.; Al-Juaid, S.S.; Irshad, A.; Warsi, M.F. Boosting the properties of Ag-decorated $Ni_3V_2O_8$ via 1D-CNTs integration for advanced photocatalytic and anti-bacterial performance. *Ceram. Int.* **2025**, *51*, 19704–19714. [CrossRef]
28. Rosalbino, F.; Macciò, D.; Scavino, G. Corrosion behaviour of Zr–Ag alloys for dental implant application. *Mater. Sci. Appl.* **2023**, *14*, 501–514. [CrossRef]
29. Lee, D.W.; Seo, D.-S. Surface Functionalization of Ag-Doped Zirconium Oxide Layers for Molecular Alignment. *FlatChem* **2025**, *50*, 100831. [CrossRef]
30. Nossova, L.; Caravaggio, G. Effect of Dopants on Soot Oxidation over Doped Ag/ZrO_2 Catalysts for Catalyzed Gasoline Particulate Filter. *Catal. Commun.* **2023**, *182*, 106744. [CrossRef]
31. Xia, C.; Zhang, N.; Chen, B.; Ma, J.; Song, T.; Wu, X.; Yang, T.; Zhang, X.; Li, Q. Tailoring Microstructure, Mechanical Properties, and Biocompatibility of Zr Alloys via the Addition of Ag. *Mater. Sci. Eng. A* **2024**, *900*, 146486. [CrossRef]
32. Nasir, N.; Rashid, M.H.; Cheema, S.A.; Rasheed, A.; Sabir, N.; Tanveer, Z.; Hassan, T.; Anjam, Q. An Experimental and Simulation Evaluation of the Structural, Morphological and Optical Characters of ZnO-Based Nano-Fibers Doped with Ag and ZrO_2. *Optik* **2022**, *265*, 169383. [CrossRef]
33. Kumar, P.; Saravanan, P.; Baskar, G.; Chitrashalini, S.; Omer, S.N.; Subashini, S.; Rajeshkannan, R.; Venkatkumar, S. Synthesis and Characterization of Ag-Decorated ZnO/MgO Nanocomposite Using a Novel Phyto-Assisted Biomimetic Approach for Anti-Microbial and Anti-Biofilm Applications. *Inorg. Chem. Commun.* **2024**, *170*, 113443. [CrossRef]
34. Khan, A.; Zaid, M.; Ameen, F.; Khan, M.A.; Kumar, S.; Al-Masri, A.A.; Islam, M.A. Colossal Antibacterial, Antibiofilm and Solar Light-Driven Photocatalytic Activity of Nanoenhanced Conjugate of Bimetallic Ag-Zr Nanoparticles with Graphene Oxide. *J. Mol. Struct.* **2024**, *1300*, 137223. [CrossRef]
35. Simon, S.M.; Prakashan, V.P.; Sajna, M.S.; Chandran, A.; George, G.; Barmiah, E.K.; Jose, G.; Biju, P.R.; Joseph, C.; Unnikrishnan, N.V. Development and Characterizations of Ag Nanoparticles Decorated TiO_2-ZrO_2 Coatings as Electrode Material for Supercapacitors. *Results Surf. Interfaces* **2023**, *10*, 100098. [CrossRef]
36. Chouhan, L.; Bouzerar, G.; Srivastava, S.K. d^0 Ferromagnetism in Ag-Doped Monoclinic ZrO_2 Compounds. *Vacuum* **2020**, *182*, 109716. [CrossRef]
37. Saleh, M.; Isik, Z.; Belibagli, P.; Arslan, H.; Gonca, S.; Özdemir, S.; Kudaibergenov, N.; Khataee, A.; Dizge, N. Fabrication of Ag Nanoparticles Coated Leonardite Basalt Ceramic Membrane with Improved Antimicrobial Properties for DNA Cleavage, *E. coli* Removal and Antibiofilm Effects. *J. Ind. Eng. Chem.* **2023**, *128*, 532–541. [CrossRef]
38. Gambardella, A.; Berni, M.; Graziani, G.; Kovtun, A.; Liscio, A.; Russo, A.; Visani, A.; Bianchi, M. Nanostructured Ag Thin Films Deposited by Pulsed Electron Ablation. *Appl. Surf. Sci.* **2019**, *475*, 917–925. [CrossRef]
39. Shang, X.; Liang, Y.; Wang, P.; Wu, Y. Synergistic Mechanism of the Fretting Wear Resistance of Fe_2O_3/Ag Nanostructured Coatings Prepared by Sliding Friction and Magnetron Sputtering. *Wear* **2024**, *546–547*, 205313. [CrossRef]
40. Zhang, P.; Zhang, X.; Hu, X.; Zhang, L.; Shi, X.; Li, Z. Simulation and Experimental Study on Frictional Wear of Plough Blades in Soil Cultivation Process Based on the Archard Model. *Biosyst. Eng.* **2024**, *248*, 190–205. [CrossRef]
41. Elahi Haghighi, N.; Hadianfard, M.J. Fabrication of Ni–ZrO_2 Nanocomposites through a New Electroforming Bath and Assessment of Their Morphology, Wear, and Corrosion Resistance. *Heliyon* **2024**, *10*, e35779. [CrossRef] [PubMed]
42. He, J.; Xie, F.; Wu, X.; Li, L.; Luo, R.; Yang, H.; Wang, S. Preparation and Friction Wear Performance of ZrO_2/MoS_2 PEO Composite Coating. *Tribol. Int.* **2025**, *202*, 110312. [CrossRef]
43. Pogodin, A.; Filep, M.; Malakhovska, T.; Vakulchak, V.; Komanicky, V.; Vorobiov, S.; Izai, V.; Shender, I.; Bilanych, V.; Kokhan, O.; et al. Recrystallization Effect on Mechanical Parameters and Increasing of Ag^+ Ionic Conductivity in $Ag_7(Si_{1-x}Ge_x)S_5I$ Ceramic Materials. *Solid State Sci.* **2023**, *140*, 107203. [CrossRef]
44. Mourya, A.K.; Gaikwad, G.S.; Singh, R.P.; Khagar, P.S.; Uke, S.J.; Wankhade, A.V. ZnO/Ag_2ZrO_3 Nanocomposites: A Tailored Nanostructure for Enhanced Supercapacitor, Photocatalytic & Antimicrobial Applications. *J. Alloys Compd.* **2024**, *1004*, 175792. [CrossRef]

45. Liu, J.; Luo, Q.; Xia, S.; Yang, X.; Lei, J.; Sun, Q.; Chen, X.; Shao, J.; Tang, X.; Zhou, G. A Cu-Ag Double-Layer Coating Strategy for Stable and Reversible Zn Metal Anodes. *J. Colloid Interface Sci.* **2024**, *665*, 163–171. [CrossRef] [PubMed]
46. Moreno Amado, M.; Alfonso, J.E.; Olaya Florez, J.J. Effect of Al and Ag Dopants on the Corrosion Resistance of the AISI 316L-YSZ System. *Ceram. Int.* **2019**, *45*, 566–572. [CrossRef]
47. Moayedee, Y.; Nikzad, L.; Fakhraei, O.; Paykar, Z.; Zekavat, E. Improvement of Mechanical, Biological, and Electrochemical Properties of Ti_6Al_4V Alloy Modified with Nb and Ag for Biomedical Applications. *J. Alloys Compd.* **2024**, *972*, 172736. [CrossRef]
48. Mina, A.; Caicedo, J.C.; Aperador, W. Sequential Analysis of Electrochemical Properties of B–Tricalcium Phosphate/Chitosan Coatings Obtained on 316L Stainless Steel. *Int. J. Electrochem. Sci.* **2013**, *8*, 11186–11200. [CrossRef]
49. Kuptsov, K.A.; Kiryukhantsev-Korneev, P.V.; Sheveyko, A.N.; Shtansky, D.V. Comparative Study of Electrochemical and Impact Wear Behavior of TiCN, TiSiCN, TiCrSiCN, and TiAlSiCN Coatings. *Surf. Coat. Technol.* **2013**, *216*, 273–281. [CrossRef]
50. Aperador, W.; Orozco-Hernández, G.; Aperador, J.; Bautista-Ruiz, J. Microstructural, Electrochemical, Mechanical, and Biocompatibility Characterization of ReN Thin Films Synthesized by DC Sputtering on Ti_6Al_4V Substrates. *Metals* **2025**, *15*, 272. [CrossRef]
51. Zhang, J.P.; He, G.; Zhu, L.Q.; Liu, M.; Pan, S.S.; Zhang, L.D. Effect of oxygen partial pressure on the structural and optical properties of ZnO film deposited by reactive sputtering. *Appl. Surf. Sci.* **2007**, *253*, 9414–9421. [CrossRef]
52. Ou, S.-F.; Chung, R.-J.; Lin, L.-H.; Chiang, Y.-C.; Huang, C.-F.; Ou, K.-L. A mechanistic study on the antibacterial behavior of silver doped bioceramic. *J. Alloys Compd.* **2015**, *629*, 362–367. [CrossRef]
53. Biju, R.F.; Jaffrin, G.; Jobisha, J.; Matharasi, A.; Prabha, S.; Vinisha, V.; Linet, M.; Mani, A.M. Structural, spectroscopic, thermal and morphological evaluation of biogenic ZnO/Ag nanocomposite using *Moringa oleifera* seed extract for enhanced antimicrobial efficacy. *Chem. Phys. Impact* **2025**, *10*, 100850. [CrossRef]

Disclaimer/Publisher's Note: The statements, opinions and data contained in all publications are solely those of the individual author(s) and contributor(s) and not of MDPI and/or the editor(s). MDPI and/or the editor(s) disclaim responsibility for any injury to people or property resulting from any ideas, methods, instructions or products referred to in the content.

coatings

MDPI

Review

Scanning Electrochemical Microscopy for Investigating Nanocomposite Epoxy Coating Degradation and Corrosion Mechanisms

Marina Samardžija [1], Marin Kurtela [2,*], Ivan Stojanović [2] and Vesna Alar [2]

1 Faculty of Mining, Geology and Petroleum Engineering, University of Zagreb, 10000 Zagreb, Croatia; marina.samardzija@rgn.unizg.hr
2 Faculty of Mechanical Engineering and Naval Architecture, University of Zagreb, 10000 Zagreb, Croatia; ivan.stojanovic@fsb.unizg.hr (I.S.); vesna.alar@fsb.unizg.hr (V.A.)
* Correspondence: marin.kurtela@fsb.unizg.hr

Abstract

Scanning Electrochemical Microscopy represents one of the most advanced high-resolution techniques that enables detailed monitoring of electrochemical processes, with a particular focus on corrosion phenomena. Scanning Electrochemical Microscopy has become an indispensable tool in studying the behavior and degradation of protective coatings exposed to aggressive environmental conditions. This technique allows researchers to precisely track local electrochemical reactions on material surfaces, providing valuable information about the stability and effectiveness of coatings. Scanning Electrochemical Microscopy enables the detection of localized current variations in the pA–nA range, allowing the identification of microdefects with nanometric width. In this paper, the basic principles of Scanning Electrochemical Microscopy operation are first presented, including a description of the device. The method of scanning the electrode is discussed through the modes and their interpretation of the obtained data for systems with protective anticorrosive coatings. Furthermore, Scanning Electrochemical Microscopy techniques enable a detailed study of the mechanisms and kinetics of new, modified coatings, which is especially significant in the case of nanoparticle-enriched coatings. Such modifications often enhance the protective properties of materials, and Scanning Electrochemical Microscopy allows monitoring of their performance under real conditions, providing insight into local electrochemical changes that are otherwise difficult to detect with standard methods. Special attention is given to the challenges researchers may encounter during experiments, such as calibration prior to measurement, interpretation of input parameters, and signal analysis. This paper aims to provide a comprehensive overview of the capabilities and limitations of Scanning Electrochemical Microscopy (SECM), emphasizing its importance as a tool for the development and optimization of new, high-performance coatings for industrial applications and scientific research.

Keywords: Scanning Electrochemical Microscopy; corrosion; coating

1. Introduction

Scanning Electrochemical Microscopy (SECM) was first introduced in 1989 independently by the research groups of Bard and Engstrom and has since developed into a versatile and powerful analytical technique. Over the past decades, Scanning Electrochemical Microscopy (SECM) has emerged as a powerful and indispensable technique

 https://doi.org/10.3390/coatings16020165

for probing interfacial processes with high spatial and temporal resolution. Its applications span a broad range of phenomena, including corrosion mechanisms, heterogeneous electron-transfer reactions, molecular transport across membranes, adsorption and desorption processes, charge transfer at liquid–liquid interfaces, and the electrochemical activity of living cells. Unlike conventional electrochemical methods, SECM combines electrochemical measurement capabilities with spatially resolved probing [1]. Electrochemical techniques such as cyclic voltammetry, alternating voltammetry, and potentiometry can be employed while simultaneously mapping local variations in electrochemical activity. By combining electrochemical analysis with scanning probe microscopy, SECM enables enhanced characterization of interfaces and local reaction dynamics. In this technique, a microelectrode—typically an ultramicroelectrode—is scanned across the surface of an immersed substrate while the resulting current response is recorded. The magnitude and spatial distribution of this current depend on both the surface topography and the electrochemical reactivity of the substrate. Through this coupling of electrochemical and positional data, SECM provides quantitative insights into heterogeneous processes at the microscale, offering a powerful platform for exploring the structure–activity relationships that govern interfacial phenomena [1–3].

To increase stability and extend the service life of epoxy coatings, many researchers are focusing their efforts on their modification. One effective approach involves incorporating nanoparticles into the epoxy coating matrix. Such modification can significantly improve barrier properties, mechanical strength, corrosion resistance, and coating durability under various operating conditions. The morphology of these systems is most analyzed using high-resolution surface techniques such as Scanning Electron Microscopy (SEM), Atomic Force Microscopy (AFM), and SECM. These methods provide detailed insight into the distribution of nanoparticles, their interaction with the polymer matrix, and surface changes caused by different types of modifications.

In their research, Zhao et al. [4] modified an epoxy coating using ZrO_2 nanoparticles. SEM analysis showed that the nanoparticles were uniformly dispersed within the epoxy matrix at lower concentrations. However, after adding 2% nano-ZrO_2 particles to the epoxy coating, agglomeration was observed, which can negatively affect the homogeneous structure and consequently reduce the mechanical and protective properties of the system. Carolan et al. [5] modified the epoxy coating with silica nanoparticles and core–shell rubber (CRS) nanoparticles. AFM analysis revealed that silica nanoparticles appear on the surface as small yellow dots, while CRS nanoparticles are identified by significantly larger dark circular structures. This difference in morphology clearly indicates the different properties and sizes of the particles used.

While SEM and AFM methods provide detailed insight into surface morphology and passive behavior, SECM analysis is used to examine the electrochemical activity and surface behavior under real conditions. This technique is capable of accurately characterizing microscopic information during the corrosion processes that are localized at specific electrochemically active sites, compared to established electrochemical measurement techniques, such as potentiodynamic polarization curve measurements, cyclic polarization, and electrochemical impedance spectroscopy (EIS), which are limited to detecting the average electrochemical information at the electrode surface [6–8].

https://doi.org/10.3390/coatings16020165

2. Principle of SECM

2.1. Instrumentation

A typical scanning electrochemical microscope comprises several key components (Figure 1). These include a low-current bipotentiostat (≤pA) for precise control and measurement of the probe and substrate potentials and currents, a high-resolution three-dimensional (3D) positioning system with stepper motors for coarse motion and piezoelectric actuators for fine adjustment, and a miniaturized probe with micrometer- to nanometer-scale dimensions (SECM tip). A data acquisition system coordinates all components to enable accurate and reliable measurements [9,10].

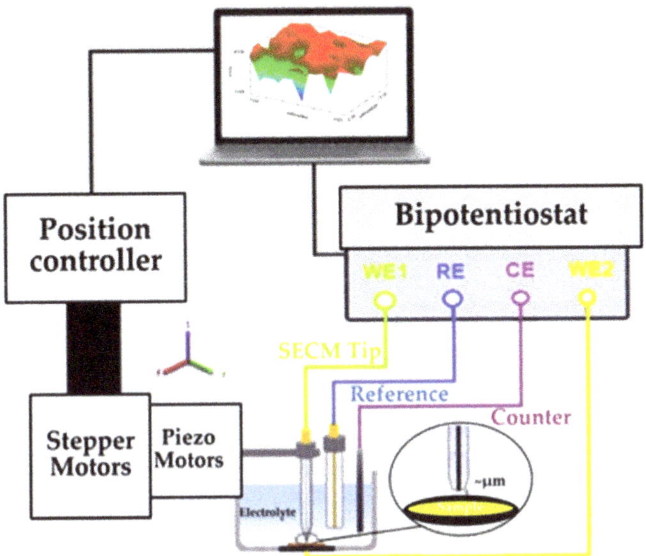

Figure 1. Diagram of the SECM apparatus [11].

The most important component of this electrochemical device is the ultramicroelectrode (UME). In the literature, these sensor elements can also be found under the name probe. At the heart of the SECM is the amperometric or potentiometric UME tip, whose position is controlled remotely, using appropriate positioners, in relation to the sample surface [12]. Their extremely small size plays a key role in the operation of the system, since the UME shows several advantages over electrodes of standard dimensions. First, the small area of the UME allows for very high sensitivity and rapid achievement of a stationary state, which results in a fast system response. In addition, these electrodes are characterized by low double-layer charging currents and minimal ohmic losses, which contribute to the stability and accuracy of measurements. The diameter of the UME directly affects the measurement resolution: the smaller the electrode, the better the resolution, and the more precise the data collected. Also, a smaller electrode area leads to an increase in current density, which results in a larger current signal and a better ability to detect weak electrochemical processes. It is important to note that reducing the dimensions of the electrode also increases sensitivity to noise, so control of instrumental and environmental factors is important for optimal system operation [10,13,14].

https://doi.org/10.3390/coatings16020165

SECM is not just another electrochemical method but represents a fundamentally different approach to conducting electrochemistry. In SECM, it is possible to apply almost all types of electrochemical measurements, including direct current (dc) and alternating current (ac) methods, and such measurements can be performed in SECM settings [15]. The device allows for the simultaneous display of two types of information: local reactivity maps (such as current distribution or resistance) and surface topography. To obtain such complementary data, combined techniques such as intermittent contact-direct current Scanning Electrochemical Microscopy (ic-dc SECM) or intermittent contact-alternating current Scanning Electrochemical Microscopy (ic-ac SECM) are used, which allow for precise differentiation of electrochemical and morphological features of the tested material [10,12].

When it is necessary to precisely define a very small reaction volume while preserving a portion of the sample unprotected from the action of the electrolyte, the Scanning Droplet System or Scanning Droplet Cell (SDS/SDC) is used. This technique uses a drop of electrolyte as a localized microreactor, which allows for selective analysis without damaging the surrounding areas. All of the above SECM methods are based on measuring local current or resistance above the surface. In contrast, the Scanning Kelvin Probe (SKP) allows for mapping the surface potential distribution, which is particularly useful in investigating corrosion processes, protective coatings, and differences in the working function potential of materials [10,15]. Table 1 lists the main SECM techniques along with the associated electrode types and their typical diameters.

Table 1. Presentation of the use of electrodes according to the SECM technique and the corresponding diameter [12,16,17].

Technique	Electrode Type	Diameter
dc-SECM	Pt/Au/C disk	1–25 μm (most often 10 μm)
ac-SECM	Pt/Au nano-/micro-	1–25 μm, nano < 500 nm
ic-dc SECM	Pt/Au disk	1–10 μm
ic-ac SECM	Pt/Au disk	1–5 μm
SDS	Micro droplet + optional UME	100–500 μm capillary/10–25 μm UME
SKP	Vibrating metal tip	50–200 μm (not UME)

2.2. Modes and Analysis of SECM Operation

The SECM device provides a series of analyses of the coating surface in the examination of corrosion processes. Surface analysis is performed using modes that indicate different modes of operation of the UME and different interactions between the UME and the sample. The original concept of SECM was based on positioning a microelectrode (tip) in close proximity to the surface of interest (the substrate) and using the flux of electrochemically active molecules as a probe to visualize the substrate and study its chemical properties [18]. Pilbáth et al. [19] conducted a study on a modified epoxy coating incorporating microcapsules of linseed oil, with the goal of enhancing its protective properties. To evaluate the electrochemical behavior of the coating after mechanical damage, they analyzed the scratched area using the dc-SECM technique. During the analysis, two operating modes of the microscope were employed: the feedback mode and the generation–collection (GC) mode. The condition of the scratched surface was examined by monitoring the oxidation of dissolved Fe^{2+} (at a tip potential of 0.5 V vs. Ag/AgCl/saturated KCl reference electrode) and the reduction of oxygen (at a tip potential of −0.7 V) in test solutions. The results showed that the oxidation of Fe^{2+} ions is not an optimal indicator of corrosion processes during prolonged exposure of the coating to the electrolyte, as the current rapidly decreases. The authors attribute this current drop to the possible deposition of corrosion products on the surface, which leads to the blocking of active sites and a reduction in

https://doi.org/10.3390/coatings16020165

the electrochemical activity of Fe^{2+} ions. Due to these limitations, Pilbáth et al. [19] proposed an alternative corrosion monitoring approach based on oxygen reduction, as this process provides more stable and reliable signals over longer periods. On the other hand, Xavier [20], in his study focusing on the use of nanoparticles in protective coatings, suggests a combined application of both methods—Fe^{2+} oxidation and oxygen reduction—in order to obtain a more comprehensive understanding of corrosion and self-healing mechanisms in the coating.

In the review paper by Nazarov and Thierry [21], coatings are analyzed using the SKP technique. SKP is a non-destructive, highly sensitive electrochemical method that measures differences in contact potential between the probe (E_{tip}) and the sample (E_{substrat}). The paper presents numerous examples of SKP applications in research, such as studying the effects of pigments, corrosion inhibitors, and surface pretreatments on coating stability, the behavior of different types of paints, and the possibility of quantitatively determining the rate of corrosion propagation beneath the coating.

The method that enables local voltametric and impedance measurements through a controlled microdroplet of electrolyte is known as SDS/SDC. This technique belongs to the group of local electrochemical methods and is used for characterizing surface properties of materials in a very limited solution volume. One of the major advantages of the SDS/SDC method is the minimal sample preparation required. Unlike conventional electrochemical techniques, no electrical contacts need to be formed on the surface, which significantly simplifies the experimental procedure and reduces the possibility of introducing artifacts. Additionally, the use of microvolumes of electrolyte limits the interaction of the solution with the broader surface area, thereby reducing potential material degradation and ensuring very low electrolyte consumption. This enables rapid, localized, and selective investigation of different points on the surface, making the method particularly useful for heterogeneous systems and materials with complex morphology [22]. In the paper by Lohrengel et al. [23], the SDS approach is described in detail, including the experimental configuration, operating principles, and a wide range of potential applications in electrochemical surface studies. A review of the available scientific literature did not reveal any studies explicitly using the SDS/SDC technique for the investigation of epoxy coatings modified with nanoparticles, indicating a promising research area that has not yet been thoroughly explored.

For contextual comparison, the principles and applications of the most used SECM method for studying nanoparticle-based coatings are presented in Table 2.

https://doi.org/10.3390/coatings16020165

Table 2. Basic information and application of SECM methods in the study of coatings for corrosion protection [21,24–37].

	SECM Operating Mode	Technique	Generator	Signal	Response	Applications
Classic Modes	Feedback mode (fb)	dc-SECM	Redox couple in solution	I_{tip}	2D current map (x-y map)	- surface reactivity assessment - homogeneity of coating and distribution of nanoparticles
		ac-SECM	AC potential (E) applied to the UME	amplitude (A), I_{tip} phase (φ)	Local |Z|(x, y), A and φ(x, y) Map of resistance (R_{ct}) and capacitance (Cdl) as a function of position	- NPs cause local changes in the current - low negative fb: improved barrier function - high fb positives: defects, uneven NP dispersion - capacitive characteristics of the coating - assessment of durability and stability of NPs in the coating
	Generation–Collection mode (G/C)	dc-SECM	E_{tip} ili I_{tip} that causes an electrochemical reaction	$I_{tip}/I_{substrat}$	Current signal as a function of time or position	- assessment of transport and barrier properties - low current: slow transport - high current: permeability, defects
	Substrate Generation/Tip Collection (SG/TC)	dc-SECM, ic-SECM,	substrate triggers an electrochemical reaction	I_{tip}	2D or 3D current map: I_{tip}(x, y) or I_{tip}(x, y, z)	- ↑ I: local anode activity - ↓ I: good barrier protection - spatial localization of defects - efficiency of NPs - assessment of sustainability of inhibition of NPs
	Tip Generation/Substrate Collection (TG/SC)	dc-SECM	UME triggers an electrochemical reaction	$I_{substrat}$	2D or 3D current map: $I_{substrat}$(x, y) or $I_{substrat}$(x, y, z)	- how quickly and how many reactive species pass through the coating to the substrate - NPs dispersion quality map (defects are seen as places of increased collector current) - a realistic picture of the durability of protection (when protective properties begin to deteriorate), where transport pathways are first opened, how NPs delay degradation - inhibition mechanism of NPs
Advance Modes	Redox Competition SECM (RC-SECM)	dc-SECM	Substrate triggers an electrochemical reaction redox mediator	I_{tip}	2D map: ΔI_{tip}(x, y) 3D map: ΔI_{tip}(x, y, z)	- uniform ↑ I_{tip}: homogeneous substrate, good dispersion of NPs, no permeable channels - local drops in the signal: a defect/pore or a thinner/weaker spot in the layer - ↑ I_{tip}: NPs act as a physical barrier - changes in the signal overtime (increase in I_{tip}: deterioration of the coating; decrease in I_{tip} overtime: activation of NPs)
	Potentiostatic/Galvanostatic (Constant-distance mode + DC feedback mode)	ic-dc SECM	depends on configuration: I_{tip}/E_{tip} or $I_{sutstrate}/E_{substrate}$	I_{tip}/E_{tip} or $I_{substrate}/E_{substrate}$	2D map: I_{tip}(x, y); E_{tip}(x, y) 3D map: I_{tip}(x, y, z); E_{tip}(x, y, z) Surface topography	- ↓ I—good barrier protection - ↑ I—defects, microcracks, poor protection - Reveals the place where the coating leaks electrolyte - Mechanism of action of NP - Influence of NP over time

https://doi.org/10.3390/coatings16020165

Table 2. *Cont.*

	SECM Operating Mode	Technique	Generator	Signal	Response	Applications
Advance Modes	Impedance mod (Constant-distance mode + AC feedback)	ic-ac SECM	AC voltage of small amplitude and specific frequency on the UME or substrate	AC through the system	2D or 3D impedance and resistant map Surface topography	- ↑ Z—better barrier protection - ↓ Z—degradation and poor NP dispersion - Heterogeneity in the signal means unequal dispersion of NPs - Monitoring of degradation over time—mechanism of NP action through active (inhibitors) or passive (barrier) protection
	Potentiometric/ Potential mapping	SKP	$E_{substrate}$ (work function)	contact potential difference (V_{CPD})	Local potential: 2D map: V(x, y) 3D map: V(x, y, z)	- shows where corrosion will start - coating homogeneity: even distribution of E - jumps of E—agglomerates, defects, weak protection - reveals hidden and initial stages of corrosion—effective NPs shift E in a more noble direction - Mapping of potentials under a thin coating—Monitoring of the "self-healing" process of the coating via shifts in E
	Feedback (DC)/ Potentiometric mode	SECM-SKP	$I_{substrate}/E_{substrate}$ (work function)	I_{tip}/V_{CPD}	2D map: I(x, y); V(x, y) 3D map: I(x, y, z); V(x, y, z)	- shows where corrosion is occurring and how much - coating homogeneity + ion permeability - shows local reactivity and corrosion kinetics - current measures, ion transport, coating permeability, nanoparticle inhibition - Monitoring the self-healing process of coatings via activation and local current decay
	Feedback/local EIS mode	SECM-SDS	UME or substrate, initiates current/impedance signal	Local current/impedance and capacitance	2D map: I(x, y); Z(x, y) 3D map: I(x, y, z); Z(x, y, z)	- simultaneous local kinetic and thermodynamic information about the coating - mapping of local corrosion activity - assessment of barrier properties of coatings - detection of micropores and defects - monitoring of distribution and efficiency of nanoparticles

https://doi.org/10.3390/coatings16020165

According to the data presented in Table 1, SECM is not only a microscopic technique, but a versatile analytical platform that can be combined with different methods to obtain different types of surface information—including chemical activity, topography, and other relevant parameters. Table 1 shows the division of SECM modes into classic and advanced. Depending on the type of experiment and the information we want to obtain, the SECM operating mode is selected. Within dc analytical methods, the device enables the application of techniques such as amperometry, voltammetry, chronoamperometry, and chronopotentiometry, which enables detailed monitoring of electroactive processes on the surface of the sample. Also, combined methods such as ic–dc SECM and ic–ac SECM represent advanced approaches that enable a deeper understanding and more precise characterization of modified surface coatings. The results of these methods can be presented as a two-dimensional (2D) image, corresponding to the operating mode known as depth scan mode. In this mode, the UME moves vertically, enabling depth scanning in the x–z or y–z plane. The depth profile is obtained by performing a series of horizontal line scans nested within the z-approach toward the sample surface. In this way, a 2D current map is generated, where the measured current is shown as a function of the UME position in the (x, z) or (y, z) coordinate system. Such a representation provides a more detailed spatial analysis of the sample's electrical properties and a clearer insight into signal variations with respect to distance and depth [38]. In 3D view, the UME scans the entire volume in three axes (x, y, z). The scan is performed by a series of 2D (x–y) maps at different z-heights. This way of displaying the results is called hopping imaging mode, XYZ Scan Mode, 3D Mapping Mode, 3D Plot Mode, or 3D SECM map [39]. These methods provide the possibility of simultaneous study of electrochemical and topographical properties, which significantly contributes to the comprehensive analysis of surface processes.

3. Overview of the Most Important SECM Methods for Studying Coatings

3.1. ic-SECM Method

ic-SECM is a completely new method that brings significant improvements compared to conventional SECM techniques. Unlike traditional SECM methods, ic-SECM allows for simultaneous monitoring of the electrochemical activity of the sample while recording its topography during measurements. The operating principle of this method is shown in Figure 2.

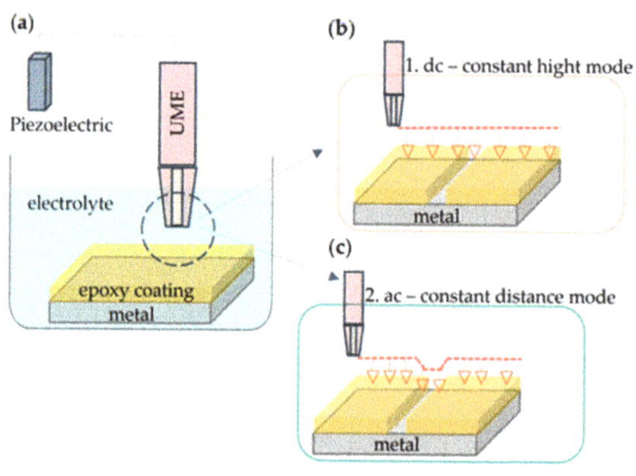

Figure 2. Schematic representation of (**a**) the electrochemical cell in the ic-SECM method, which operates based on two principles: (**b**) using direct current and (**c**) alternating current.

https://doi.org/10.3390/coatings16020165

As shown in Figure 2a, the distance of the UME, which is located in the electrolyte, from the sample is controlled using a piezoelectric positioner controller. The piezo attachment allows for fine positioning of the UME very close to the sample surface and control of the distance between the electrode and the surface [40]. Likewise, in some modes (such as ic-SECM) the piezo can vibrate or measure the probe for contact control or distance-controlled imaging. The ic-SECM method operates on two principles, depending on the type of electrochemical signal that we want to measure in interaction with the sample. If we want to observe current changes across the sample surface, we use a direct current (DC) signal in combination with ic-SECM. In this approach, a piezo controller maintains a constant distance between the probe and the sample surface. As shown in Figure 2b, the topography is determined by the distance of the probe from the sample [41]. In contrast, if we want to observe impedance responses across the sample surface, an alternating current (AC) signal is used in combination with ic-SECM. In this method, the piezo controller allows the probe to oscillate relative to the sample surface. As shown in Figure 2c, the probe remains at a constant distance from the sample, meaning that the probe height follows the topographic changes [41].

In order to precisely position the UME above the sample surface before performing in situ SECM measurements, it is first necessary to perform a DC piezo-calibration in the range from nanometers to micrometers. DC piezo measurement enables very precise movement of the probe in all three axes (x, y, z) and serves to determine the distance between the tip of the UME and the surface of the sample. In Figure 3a, piezo DC response shows the dependence between the applied voltage and the actual position of the UME. The results show the occurrence of hysteresis, because the piezo actuator does not follow the same path during voltage increase and decrease. This phenomenon can lead to scanning errors, especially at high resolutions, if the hysteresis is not properly compensated through software or hardware correction. Before performing the in situ SECM measurement, it is necessary to perform the AC piezo-calibration. This step makes it possible to determine the optimal frequency and amplitude of oscillations of the piezo element, which ensures a stable and linear response during the measurement. In Figure 3b, the dependence of signal gain at frequency on frequency is shown, which is used to select the frequency at which the system response is most reliable and with minimal signal distortion. Before each measurement, it is necessary to properly adjust the frequency at which the probe oscillations will be stable and regular, using a piezo controller [42,43].

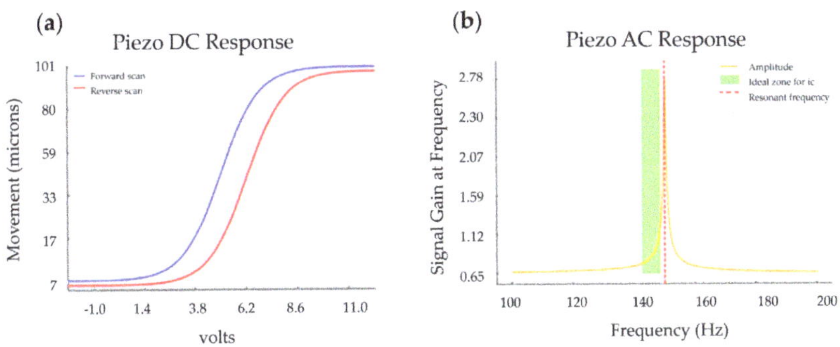

Figure 3. Calibration measurement of (**a**) piezo DC response and (**b**) piezo AC response.

The green zone in Figure 3b indicates the ideal frequency zone for the ic method, usually close to, but slightly below, resonance (≈95–99%). The ideal frequency indicates that the oscillations will be more efficient, which means that maximum sensitivity and stability of the contact interaction with the surface will be obtained. If the frequency of

https://doi.org/10.3390/coatings16020165

the UME in the intermittent contact mode is not optimally selected, various problems may occur, depending on whether it is too low or too high (Table 3).

Table 3. UME response to frequencies below and above resonant [44,45].

Frequency	Problems	Consequences
too low frequency	UME oscillates weakly and may touch the surface for too long	- Surface damage - Increased UME consumption - Noise in the signal
too high frequency	UME does not follow resonance and probe jumping may occur.	- Loss of sensitivity - Noisy and unreliable signal - Data may be misinterpreted.

3.2. dc-SECM Method for Investigation Coating

3.2.1. Principles of ic-dc SECM Technique

A direct current bias is applied to the UME, which is positioned in close proximity to the sample. This bias interacts with the sample via a redox mediator. The current measured by the probe depends on the interaction of the mediator row with the sample and reflects the relative electrochemical activity of the sample in the area immediately below the probe [46]. This method is most used for the analysis of coatings containing nanoparticles, with the aim of gaining a better understanding of their electrochemical mechanism of action and corrosion protection properties. Before the measurement begins, a pre-prepared protective coating is applied to a 1 cm × 1 cm metal substrate. After a certain drying period—depending on the composition of the coating and environmental conditions—a small scratch is carefully made on the surface of the sample to simulate mechanical damage or initial defects. Depending on the type of mediator to be observed, an appropriate electrochemical solution is selected. The most used solution for testing protective coatings is an aqueous solution of 3.5% NaCl, as sodium chloride is one of the most aggressive components that can accelerate coating degradation and promote corrosion processes. However, during UME measurements, NaCl does not participate directly in the electrochemical reaction due to the applied potential. Under such conditions, the reaction occurs exclusively with the aqueous solution or the metallic substrate. In this method, a constant potential is applied to the UME, which can convert the redox mediator species present in the electrolyte solution [25]. If the reduction of dissolved oxygen above both the undamaged and damaged surface of the sample is to be observed, a potential of −0.7 V vs. AgCl is applied to the UME. A schematic of this measurement is shown in Figure 4, where the reduction of oxygen and iron dissolution on the epoxy surface were observed.

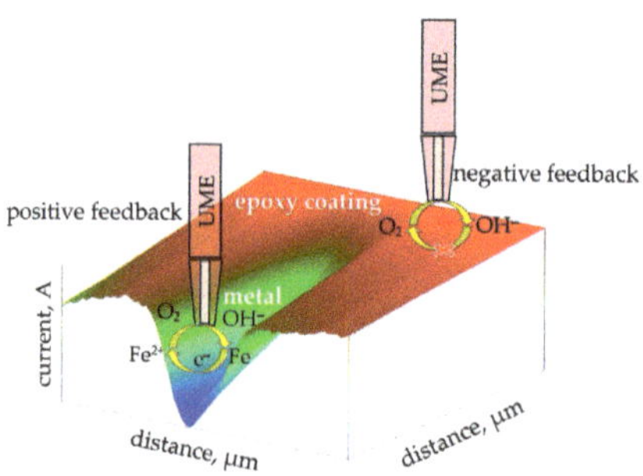

Figure 4. Schematic representation of positive and negative feedback on a scratched epoxy substrate during ic-dc SECM measurements.

https://doi.org/10.3390/coatings16020165

The red areas in Figure 4 represent the undamaged epoxy coating. The epoxy coating is an insulating material, which results in a negative feedback signal during measurement, characteristic of the amperometric feedback mode (U = constant, I $\neq$ constant). At the very tip of the UME, the reduction of dissolved oxygen from the aqueous solution to hydroxide ions (OH^-) occurs, according to the following electrochemical reaction:

$$O_2(g) + 4e^- + 2H_2O(l) \rightarrow 4OH^-(aq) \quad (1)$$

However, due to the presence of the insulating epoxy coating, the resulting OH^- ions cannot be regenerated, and their diffusion towards the tip of the UME is blocked. As a result, there is a decrease in electrical current and an increase in surface resistance, which confirms the undamaged nature of the coating. On the other hand, in the damaged areas, where the UME comes into contact with the conductive metal substrate, a positive feedback signal is observed. Immersing such a surface in the electrolyte creates potential at the metal/solution interface, which, according to the Nernst equation, depends on the concentration of redox species [25]. In such zones, electrochemical dissolution of iron occurs according to the following reaction:

$$Fe(s) \rightarrow Fe^{2+}(aq) + 2e^-. \quad (2)$$

The electrons released during this process can be accepted by dissolved oxygen from the aqueous solution. The resulting oxidation product, O_2, diffuses towards the surface of the UME, where it can be further reduced to OH^- ions. This phenomenon allows the maintenance of electrochemical activity at the damaged areas, which is reflected by an increase in current and a decrease in resistance compared to the undamaged areas covered by epoxy.

In the review paper by Santana et al. [47], a detailed list of redox mediators used for the examination of various types of coatings on a wide range of materials is provided. The authors provide an overview of the application of specific mediators depending on the type of coating, substrate, and the purpose of the investigation, thereby enabling a better selection of experimental conditions. Coating tests are most often conducted in a simulated corrosive aqueous sodium chloride (NaCl) solution, which is used as a model electrolyte due to its similarity to real corrosive environments. From the tables presented in the study [47], key electrochemical parameters necessary for the proper execution of measurements can be identified, including working potentials, the choice of redox pairs, and optimal experimental conditions. In the case of measuring the concentration of dissolved oxygen (O_2) in NaCl solution, an ultramicroelectrode (UME) is used, whose tip is polarized at a potential of −0.70 V relative to an Ag/AgCl/KCl (saturated) reference electrode. At this potential, oxygen is reduced, allowing quantitative monitoring of its concentration in the solution, and thereby assessing the protective properties of the tested coating [47].

When observing the current of iron oxidation on the damaged epoxy surface, the principle is the same, but the potential on the UME is typically around +0.60 V vs. Ag/AgCl.

3.2.2. ic-dc SECM for Investigation of Coating Protection

Traditional methods have long been employed to evaluate protective coatings, but advances in modern coating technologies now demand more sophisticated characterization techniques. Conventional approaches are insufficient to explain the behavior of nanoscale pigments (20–30 nm) or responsive "smart" coatings [48]. In corrosion science, SECM has proven to be a powerful tool for investigating both organic and inorganic metal

https://doi.org/10.3390/coatings16020165

coatings, enabling applications such as microscopic chemical imaging, measurement of physicochemical parameters, and micromachining.

Xavier conducted extensive research on the role of nanoparticles (TiO_2, ZrO_2, Mg, Ce) in reducing corrosion in epoxy coatings exposed to saltwater [20,49,50]. In the study, a platinum microelectrode was scanned across damaged areas of the coating to monitor dissolved oxygen through the cathodic current. Because oxygen reduction dominates the cathodic corrosion process, changes in dissolved oxygen concentration served as the primary indicator of corrosion activity. The results demonstrated that all nanoparticles reduced the current over damaged regions compared to coatings without nanoparticles, indicating that their presence suppresses iron dissolution from the metal surface. This protective effect is attributed to the formation of surface complexes, which provide additional barrier protection beyond the coating's inherent film resistance. Madhankumar et al. [51] investigated the behavior of Si nanoparticles (Si NPs) in scratched epoxy-coated samples. SEM and EDS analyses revealed that Si NPs migrated from the coating to the damaged areas over time. It is proposed that Si dissolved as ions (Si^{n+}) and subsequently redeposited at the scratches, providing localized corrosion protection. This behavior indicates a sacrificial effect of Si, enhancing the coating's ability to protect the underlying steel. Furthermore, Xia et al. [52] examined coating defects using cyclic voltammetry with ferrocenylmethanol as a redox mediator. SECM analysis revealed that defect activity initially increased, reflecting rapid early-stage corrosion of the carbon steel, and later decreased due to the formation of protective corrosion products. Modern SECM instruments allow for the visualization of results in a 3D format, as demonstrated by Kurtela et al. [53]. In their study, they presented 3D results obtained using the ic-dc SECM technique, examining the current distribution on damaged areas of coatings made of pure epoxy (Figure 5a) and coatings modified with graphene oxide (GO) and aluminum nanoparticles (Al NPs) (Figure 5b). This 3D representation clearly shows that higher current densities are observed on the pure epoxy coating, indicating a more pronounced Fe^{2+} dissolution at the scratch site. In contrast, the presence of GO and Al NP in the coating significantly reduces this process, suggesting improved corrosion protection in the modified samples.

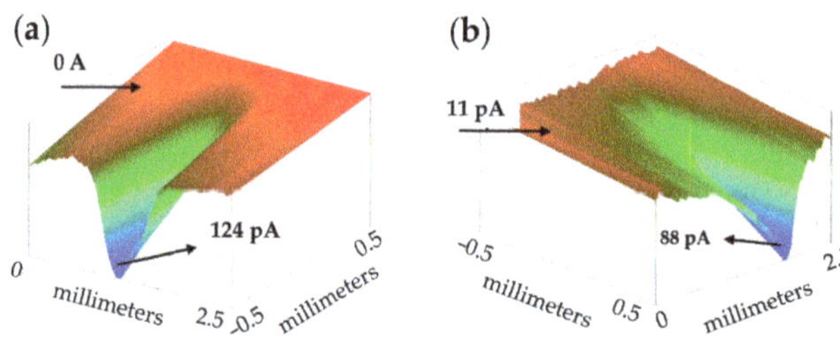

Figure 5. ic-dc SECM distribution of current of scratched samples of (**a**) epoxy coating and (**b**) EP + Al NP + GO sample with cross section immediately after immersed in 3.5 wt% NaCl solution at the tip potential of +60 V vs. Ag/AgCl/KCl reference electrode [53].

3.3. ic-ac-SECM Method for Investigation Coating

For the ic-ac-SECM method, as shown in Figure 2c, it is characteristic that the UME signal, to which the potentiometric method (U $\neq$ constant) is applied, depends on the distance from the sample. The ic-ac-SECM method provides information about both the topography and electrochemical characteristics of the surface, in contrast to the ac-SECM method, which provides information about the local electroactivity of the sample.

https://doi.org/10.3390/coatings16020165

The advantage of using the ac-SECM method over the dc-SECM method lies in the elimination of the need for a redox mediator. While the selection of an appropriate redox mediator is crucial in dc-SECM, special attention must be given when investigating corrosion processes. This is because different electrolytes can significantly affect the outcome of the examination—they may accelerate or inhibit the corrosion process, which can lead to incorrect interpretations of results. In this context, the ac-SECM method presents a superior alternative, as it enables more accurate and reliable measurements without the influence of the redox mediator on the system.

3.3.1. Principles of ac-SECM Technique

Measurements in the ac-SECM method require the use of a UME, reference electrode, working electrode, and auxiliary electrode. All three electrodes are immersed in the electrolyte, with deionized water being most used due to its low conductivity (Figure 6). The lateral electrochemical contrast in ac-SECM has been found to be highly dependent on the frequency of the applied alternating voltage [54]. An alternating voltage of a specific frequency is applied to the UME, resulting in the appearance of alternating current at the tip of the UME. The current pathway depends on the distance between the UME and the substrate, as well as the type of substrate—whether it is an insulating or conductive surface.

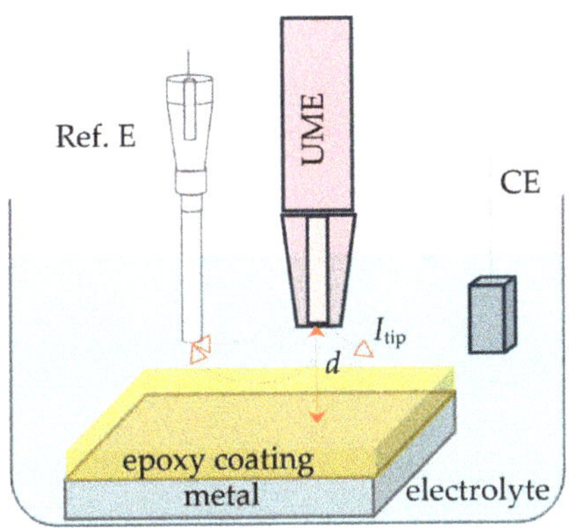

Figure 6. Schematic representation of ac-SECM technique.

In the case where the distance (d) between the UME and the sample is too large, the current will flow through the solution without interacting with the sample, meaning it will not be directed locally towards the sample. The electric field will extend throughout the entire volume of the solution, and as a result, bulk impedance will be measured.

When the UME is at a short distance from the sample, interaction occurs depending on the material of the sample. Figure 7 illustrates the ac-SECM equivalent circuits that describe the overall impedance response on a surface made of insulating and conductive materials. If the sample is an insulator (Figure 7a), the current, I_{tip}, encounters the blocking nature of the sample surface, resulting in a high resistance denoted as Z_{surf} [55]. Under such conditions, the behavior of the electric circuit can be described by a simple Randles circuit, where the decrease in current leads to an increase in resistance. Since I_{tip} cannot pass through the insulating sample, it is redirected towards the reference electrode.

https://doi.org/10.3390/coatings16020165

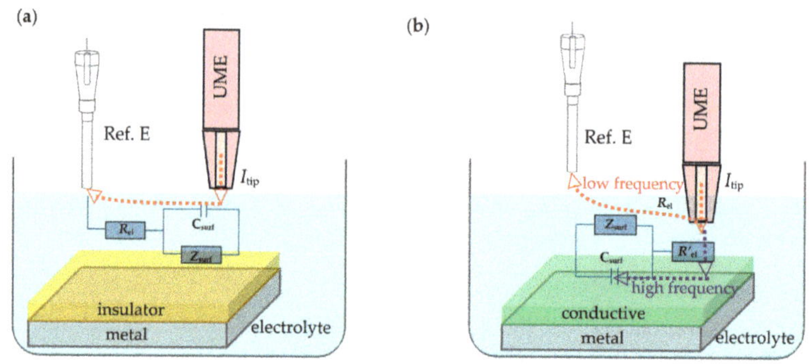

Figure 7. Scheme of ac-SECM equivalent circuits on a surface made of (**a**) insulating and (**b**) conducting materials.

In the application of the ac-SECM technique, measurements on conductive substrates can be performed at both high and low frequencies. At high frequencies, when the sample is conductive (Figure 7b), the tip current (I_{tip}) flowing from the UME is limited by the resistance of the electrolyte between the UME tip and the sample surface, denoted as (R'_e) [56]. In this regime, the current primarily flows capacitively, through charge displacement. This capacitive component of the current is particularly important when studying thin coatings, as it allows for detailed monitoring of changes in layer thickness, porosity, the presence of cracks, and overall uniformity of the coating. In other words, high-frequency measurements provide insight into the physical structure and integrity of the layer without direct contact with the electrically conductive sample, which is crucial for analyzing coatings that are sensitive to mechanical damage or corrosion [47,57].

Unlike measurements at high frequencies, where the current is dominated by accumulation in the electrical double layer, at low frequencies the current primarily depends on the properties of the electrolyte and enters the substrate through an electrochemical reaction, i.e., as faradaic current. In this regime, the current no longer accumulates in the double layer but instead directly follows the kinetics of the electrochemical reactions and the ionic conductivity of the solution [25,42]. In Figure 7b, the I_{tip} current component travels from the top of the UME toward the substrate, where charge transfer occurs, and then spreads further toward the reference electrode, completing the current loop. Such current behavior is characteristic of systems in which faradaic processes dominate over capacitive contributions. In low-frequency measurements, an increase in Faradaic current can indicate the presence of localized corrosion processes beneath a protective coating. The magnitude of this current provides a quantitative insight into the kinetics of metal redox reactions, enabling more precise monitoring of corrosion rates and identification of critical areas particularly susceptible to degradation. Such data are essential for optimizing protective coatings and predicting the service life of metallic structures in aggressive environmental conditions [25,34,54].

3.3.2. ic-ac-SECM for Investigation of Coating Protection

Eckhard et al. [54] conducted ac-SECM testing on a polymer coating sample with defects, over a wide frequency range from 0.273 Hz to 33.199 kHz. The aim of the study was to determine the optimal parameters for achieving high image resolution, thereby improving the accuracy of defect detection and characterization within the coating. In their study, it is observed that as the frequency increases, the defect becomes more confined.

Eckhard et al. [54] report that the surface response investigated at high perturbation frequencies is significantly more constrained compared to lower frequencies, at which

https://doi.org/10.3390/coatings16020165

capacitive contributions also influence the overall electrochemical signal. Consequently, scratches in the images at frequencies below 1 kHz appear considerably broader.

Wang et al. [58] investigated scratched and self-healed coatings using AC-SECM at both high and low perturbation frequencies. They showed that the AC-SECM images were highly dependent on the probing frequency and the selected portion of the impedance response. At a high frequency (60,000 Hz), the scratch topography was clearly resolved, with minimal influence from surrounding areas. At a low frequency (200 Hz), the scratch remained visible, but the surrounding region was also probed, reducing spatial resolution. An intermediate frequency (3464 Hz) produced results similar to those at high frequency. These findings highlight AC-SECM's ability to capture a wide range of local electrochemical topographies, enabling detailed observation of self-healing processes at different frequencies.

Unlike the previous two authors, Samardžija et al. [11] analyze the influence Al NPs on the anticorrosion properties of an epoxy coating in their study. Their results indicate that the addition of nanoparticles significantly improves the barrier properties of the coating and slows down the penetration of the corrosive medium. Experimental data for the epoxy sample with 1% Al NP content are shown in Figure 8. Figure 8a presents the Bode plot, which clearly shows that the nanocomposite coating exhibits a higher overall resistance and provides better protection compared to the pure epoxy coating after 24 h of exposure to a 3.5% NaCl solution. The authors further expand their analysis of the Bode plots by examining the capacitive properties of the coating at a frequency of 100 kΩ using the ic-ac SECM technique, which enables local mapping of the electrochemical properties of the surface. Figure 8b shows the spatial distribution of impedance values after 1 h of exposure to the solution, while Figure 8c shows the same distribution after 30 days of immersion in the 3.5% NaCl solution. Based on these results, it can be concluded that although the overall impedance values decrease over time due to the gradual degradation process, areas of elevated impedance simultaneously expand. This phenomenon is associated with the formation of an aluminum oxide film on the surface of the nanocomposite, which contributes to additional passivation and slows down the corrosion process.

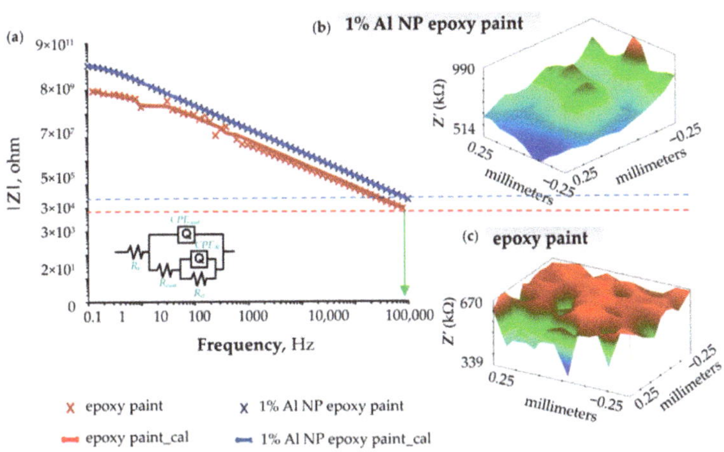

Figure 8. Representation of (**a**) Bode diagram of epoxy coating and 1% Al NP epoxy nanocomposite after 1 h exposure to a corrosive medium, (**b**) surface topography of a nanocomposite with 1% Al NPs after 1 h exposure to a corrosive medium, (**c**) surface topography of a nanocomposite with 1% Al NPs after 30 days of exposure to a corrosive medium [11].

https://doi.org/10.3390/coatings16020165

4. Conclusions

A review of the available literature confirms that SECM is one of the key and most promising techniques for studying degradation and corrosion processes in epoxy coatings. This paper presents a concise yet comprehensive overview of SECM techniques currently used to examine and interpret the mechanisms of anticorrosion protection in modified coatings. By analyzing various SECM modes and their experimental approaches, the paper provides a clear and structured insight into the acquisition, processing, and interpretation of SECM data.

Through the review of numerous studies, it is emphasized that the incorporation of nanoparticles such as ZrO_2, TiO_2, CeO_2, Mg, Al, SiO_2, or graphene oxide (GO) can significantly enhance the barrier properties of coatings, reduce Fe^{2+} dissolution from the metal substrate, and promote self-healing or passivation processes. Such modifications slow down electrolyte penetration into the coating and contribute to long-term stability and protective performance.

Special attention is given to the importance of proper calibration, with detailed explanations of DC and AC piezo-calibration procedures. The paper also clarifies the critical role of frequency optimization due to its strong influence on image resolution, as well as the challenges that arise at excessively low or high frequencies. This highlights the need for precisely determining the optimal operating conditions of the UME probe to ensure reliability and reproducibility of the acquired data.

Overall, this work provides both a theoretical framework and practical guidelines for the application of SECM in the field of metal protection, clearly demonstrating that this technique enables insight into corrosion processes at spatial and temporal scales beyond the capabilities of standard analytical methods. SECM thus stands out as a highly valuable methodology for the development of advanced, durable, and functional protective coatings.

Author Contributions: Conceptualization, M.S. and V.A.; methodology, M.S. and M.K.; software, M.S. and M.K.; validation, I.S., V.A. and M.K.; formal analysis, M.S.; investigation, M.S., M.K. and I.S.; resources, M.S., M.K. and I.S.; data curation, M.S.; writing—original draft preparation, M.S.; writing—review and editing, M.S.; visualization, M.S.; supervision, V.A., I.S. and M.K.; project administration, V.A.; funding acquisition, V.A. All authors have read and agreed to the published version of the manuscript.

Funding: This research received no external funding.

Institutional Review Board Statement: Not applicable.

Informed Consent Statement: Not applicable.

Data Availability Statement: The data presented in this study are available on request from the corresponding author.

Conflicts of Interest: The authors declare no conflicts of interest.

References

1. Polcari, D.; Dauphin-Ducharme, P.; Mauzeroll, J. Scanning electrochemical microscopy: A comprehensive review of experimental parameters from 1989 to 2015. *Chem. Rev.* **2016**, *116*, 13234–13278. [CrossRef]
2. Lillehei, P.T.; Bottomley, L.A. Scanning probe microscopy. *Anal. Chem.* **2000**, *72*, 189R–196R. [CrossRef]
3. Poggi, M.A.; Bottomley, L.A.; Lillehei, P.T. Scanning microscopy. *Anal. Chem.* **2002**, *74*, 2851–2862. [CrossRef]
4. Zhao, X.; Liu, S.; Wang, X.; Hou, B. Surface modification of ZrO_2 nanoparticles with styrene coupling agent and its effect on the corrosion behaviour of epoxy coating. *Chin. J. Oceanol. Limnol.* **2014**, *32*, 1163–1171. [CrossRef]
5. Carolan, D.; Ivankovic, A.; Kinloch, A.J.; Sprenger, S.; Taylor, A.C. Toughening of epoxy-based hybrid nanocomposites. *Polymer* **2016**, *97*, 179–190. [CrossRef]
6. Invernizzi, A.J.; Sivieri, E.; Trasatti, S.P. Corrosion behaviour of Duplex stainless steels in organic acid aqueous solutions. *Mater. Sci. Eng. A* **2008**, *485*, 234–242. [CrossRef]

https://doi.org/10.3390/coatings16020165

7. Martins, M.; Casteletti, L.C. Microstructural characteristics and corrosion behavior of a super duplex stainless steel casting. *Mat. Charact.* **2009**, *60*, 150–155. [CrossRef]
8. Lo, I.H.; Tsai, W.T. Effect of selective dissolution on fatigue crack initiation in 2205 duplex stainless steel. *Corrosi. Sci.* **2007**, *49*, 1847–1861. [CrossRef]
9. Engstrom, R.C.; Wightman, R.M.; Kristensen, E.W. Diffusional distortion in the monitoring of dynamic events. *Anal. Chem.* **1988**, *60*, 652–656. [CrossRef]
10. Bard, A.J.; Faulkner, L.R. *Electrochemical Methods: Fundamentals and Applications*, 2nd ed.; John Wiley & Sons Inc.: New York, NY, USA, 2001.
11. Samardžija, M.; Stojanović, I.; Kurtela, M.; Alar, V. Influence of aluminum nanoparticles in epoxy resin and epoxy coating for anticorrosion and antibacterial protection in pipeline industry. *J. Appl. Polym. Sci.* **2024**, *141*, e55002. [CrossRef]
12. Edwards, M.A.; Martin, S.; Whitworth, A.L.; Macpherson, J.V.; Unwin, P.R. Scanning electrochemical microscopy: Principles and applications to biophysical systems. *Physiol. Meas.* **2006**, *27*, R63–R108. [CrossRef]
13. Dowell, T.J.; Hewage, J.; Wipf, D.O. How to prepare an ultramicroelectrode with a precise geometry. *ChemRxiv* **2025**, 1–19. [CrossRef]
14. Basha, C.A.; Rajeddran, L. Theories of ultramicrodisc Eelectrodes. *Int. J. Electrochem. Sci.* **2006**, *1*, 268–282. [CrossRef]
15. Mirkin, M.V.; Horrocks, B.R. Electroanalytical measurements using the scanning electrochemical microscope. *Anal. Chim. Acta* **2000**, *406*, 119–146. [CrossRef]
16. Huang, L.; Li, Z.; Lou, Y.; Cao, F.; Zhang, D.; Li, X. Recent advances in scanning electrochemical microscopy for biological applications. *Materials* **2018**, *11*, 1389. [CrossRef]
17. Morkvenaite-Vilkonciene, I.; Ramanaviciene, A.; Kisieliute, A.; Bucinskas, V.; Ramanavicius, A. Scanning electrochemical microscopy in the development of enzymatic sensors and immunosensors. *Biosens. Bioelectron.* **2019**, *141*, 111411. [CrossRef] [PubMed]
18. Mirkin, M.V.; Nogala, W.; Velmurugan, J.; Wang, Y. Scanning electrochemical microscopy in the 21st century. Update 1: Five years after. *Phys. Chem. Chem. Phys.* **2011**, *13*, 21196–21212. [CrossRef]
19. Pilbáth, A.; Szabó, T.; Telegdi, J.; Nyikos, L. SECM study of steel corrosion under scratched microencapsulated epoxy resin. *Prog. Org. Coat.* **2012**, *75*, 480–485. [CrossRef]
20. Xavier, J.R.; Nallaiyan, R. Application of EIS and SECM studies for investigation of anticorrosion properties of epoxy coatings containing ZrO_2 nanoparticles on mild steel in 3.5% NaCl solution. *J. Fail. Anal. Prev.* **2016**, *16*, 1082–1091. [CrossRef]
21. Nazarov, A.; Thierry, D. Application of Scanning Kelvin Probe in the study of protective paints. *Front. Mater.* **2019**, *6*, 192. [CrossRef]
22. Joress, H.; DeCost, B.; Hassan, N.; Braun, T.M.; Gorham, J.M.; Hattrick-Simpers, J. Development of an automated millifluidic platform and data-analysis pipeline for rapid electrochemical corrosion measurements: A pH study on Zn-Ni. *Electrochim. Acta* **2022**, *428*, 140866. [CrossRef]
23. Lohrengel, M.M.; Moehring, A.; Pilaski, M. Electrochemical surface analysis with the scanning droplet cell. *Fresenius J. Anal. Chem.* **2000**, *367*, 334–339. [CrossRef]
24. Bard, A.J.; Crayston, J.A.; Kittlesen, G.P.; Vareo Shea, T.; Wrighton, M.S. Digital Simulation of the measured electrochemical response of reversible redox couples at microelectrode arrays: Consequences arising from closely spaced ultramicroelectrodes. *Anal. Chem.* **1986**, *58*, 2321–2331. [CrossRef]
25. Eckhard, K.; Schuhmann, W. Alternating current techniques in scanning electrochemical microscopy (AC-SECM). *Analyst* **2008**, *133*, 1486–1497. [CrossRef] [PubMed]
26. Fayyad, E.M.; Almaadeed, M.A.; Jones, A.; Abdullah, A.M. Evaluation techniques for the corrosion resistance of self-healing coatings. *Int. J. Electrochem. Sci.* **2014**, *9*, 4989–5011. [CrossRef]
27. González-García, Y.; Santana, J.J.; González-Guzmán, J.; Izquierdo, J.; González, S.; Souto, R.M. Scanning electrochemical microscopy for the investigation of localized degradation processes in coated metals. *Prog. Org. Coat.* **2010**, *69*, 110–117. [CrossRef]
28. Kellenberger, A.; Vaszilcsin, N.; Dan, M.L.; Mitelea, I.; Geana, A.A.; Lædre, S.; Craciunescu, C.M. Corrosion investigation by Scanning electrochemical microscopy of AISI 446 and Ti-Coated AISI 446 ferritic stainless steel as potential material for bipolar plate in PEMWE. *Acta Metall. Sin.* **2024**, *37*, 607–619. [CrossRef]
29. Nonnenmacher, M.; O'Boyle, M.P.; Wickramasinghe, H.K. Kelvin probe force microscopy. *Appl. Phys. Lett.* **1991**, *58*, 2921–2923. [CrossRef]
30. Maya-Cornejo, J.; Rodríguez-Gómez, F.J.; Molina, G.A.; Galindo-de-la-Rosa, J.; Ledesma-García, J.; Hernández-Martínez, A.R.; Esparza, R.; Estévez, M. Electrochemical study of a hybrid polymethyl methacrylate coating using SiO_2 nanoparticles toward the mitigation of the corrosion in marine environments. *Materials* **2019**, *12*, 3216. [CrossRef]
31. Santana Santos, C.; Jaato, B.N.; Sanjuan, I.; Schuhmann, W.; Andronescu, C. Operando Scanning electrochemical probe microscopy during electrocatalysis. *Chem. Rev.* **2023**, *123*, 4972–5091. [CrossRef]

https://doi.org/10.3390/coatings16020165

32. Raj, X.J. Application of EIS and SECM studies for investigation of anticorrosion properties of epoxy coatings containing zinc oxide nanoparticles on mild steel in 3.5% NaCl solution. *J. Mater. Eng. Perform.* **2017**, *26*, 3245–3253. [CrossRef]
33. Souto, R.M.; Lamaka, S.; González, S. Uses of Scanning electrochemical microscopy in corrosion research. In *Microscopy: Science, Technology, Applications and Education*, 2nd ed.; Méndez-Vilas, A., Díaz, J., Eds.; Formatex: Badajoz, Spain, 2010; Volume 3, pp. 1769–1780.
34. Traxler, I.; Singewald, T.D.; Schimo-Aichhorn, G.; Hild, S.; Valtiner, M. Scanning electrochemical microscopy methods (SECM) and ion-selective microelectrodes for corrosion studies. *Corros. Rev.* **2022**, *40*, 515–542. [CrossRef]
35. Trentin, A.; Pakseresht, A.; Duran, A.; Castro, Y.; Galusek, D. Electrochemical characterization of polymeric coatings for corrosion protection: A review of advances and perspectives. *Polymers* **2022**, *14*, 2306. [CrossRef]
36. Udhayakumar, M.; Radhika, N.; Arun, K.L. A comprehensive review on nanocrystalline coatings: Properties, challenges and applications. *J. Bio. Tribo-Corros.* **2022**, *8*, 83. [CrossRef]
37. Upadhyay, V.; Battocchi, D. Localized electrochemical characterization of organic coatings: A brief review. *Prog. Org. Coat.* **2016**, *99*, 365–377. [CrossRef]
38. Filice, F.P.; Li, M.S.M.; Wong, J.; Ding, Z. SECM stack imaging and full 3D modelling of asymmetric clusters of live cells. *Electrochim. Acta* **2025**, *538*, 146981. [CrossRef]
39. Lazenby, R.A.; McKelvey, K.; Unwin, P.R. Hopping intermittent contact-scanning electrochemical microscopy (HIC-SECM): Visualizing interfacial reactions and fluxes from surfaces to bulk solution. *Anal. Chem.* **2013**, *85*, 2937–2944. [CrossRef]
40. Putnam, S.T.; Santiago-Carboney, A.; Qian, P.; Rodríguez-López, J. Scanning Electrochemical Microscopy (SECM): An evolving toolbox for revealing the chemistry within electrochemical processes. *Anal. Chem.* **2025**, *97*, 8147–8181. [CrossRef] [PubMed]
41. McKelvey, K.; Edwards, M.A.; Unwin, P.R. Intermittent contact-scanning electrochemical microscopy (IC-SECM): A new approach for tip positioning and simultaneous imaging of interfacial topography and activity. *Anal. Chem.* **2010**, *82*, 6334–6337. [CrossRef]
42. Bard, A.J.; Mirkin, M.V. *Scanning Electrochemical Microscopy*, 2nd ed.; CRC Press: New York, NY, USA, 2012.
43. Wipf, D.; Bard, A.J. Scanning electrochemical microscopy. 15. Improvements in imaging via tip-position modulation and lock-in detection. *Anal. Chem.* **1992**, *64*, 1362–1367. [CrossRef]
44. Deng, W. Cantilever Dynamic Analysis in Tapping Mode Atomic Force Microscopy. Ph.D. Thesis, Liverpool John Moores University, Liverpool, UK, April 2015.
45. Keyvani, A.; Tamer, M.S.; van Wingerden, J.W.; Goosen, J.F.L.; van Keulen, F. A comprehensive model for transient behavior of tapping mode atomic force microscope. *Nonlinear Dyn.* **2019**, *97*, 1601–1617. [CrossRef]
46. Catarelli, S. Scanning electrochemical microscopy: A local view of electrochemistry. *Infocus* **2020**, *57*. [CrossRef]
47. Santana, J.J.; Izquierdo, J.; Souto, R.M. Uses of Scanning electrochemical microscopy (SECM) for the characterization with spatial and chemical resolution of thin surface layers and coating systems applied on metals: A review. *Coatings* **2022**, *12*, 637. [CrossRef]
48. Farrokhpay, S. Application of spectroscopy and microscopy techniques in surface coatings evaluation: A review. *Appl. Spectrosc. Rev.* **2012**, *47*, 233–243. [CrossRef]
49. Xavier, J.R.; Nishimura, T. Evaluation of the corrosion protection performance of epoxy coatings containing Mg nanoparticle on carbon steel in 0.1 M NaCl solution by SECM and EIS techniques. *J. Coat. Technol. Res.* **2017**, *14*, 395–406. [CrossRef]
50. Raj, X.J. Electrochemical investigation into the effect of nano-titania on the protective properties of epoxy coatings on mild steel in natural seawater. *Int. Petrochem. Sci. Eng.* **2017**, *2*, 29–37.
51. Madhankumar, A.; Rajendran, N.; Nishimura, T. Influence of Si nanoparticles on the electrochemical behavior of organic coatings on carbon steel in chloride environment. *J. Coat. Technol. Res.* **2012**, *9*, 609–620. [CrossRef]
52. Xia, D.H.; Wang, J.; Wu, Z.; Qin, Z.; Xu, L.; Hu, W.; Behnamian, Y.; Luo, J.L. Sensing corrosion within an artificial defect in organic coating using SECM. *Sens. Actuators. B. Chem.* **2019**, *280*, 235–242. [CrossRef]
53. Kurtela, M.; Samardžija, M.; Bujak, M.; Stojanović, I.; Bojanić, K.; Ljubek, G. Investigation of corrosion resistance of epoxy coatings reinforced with graphene oxide and aluminium nanoparticle. *Rud. Geol. Naft. Zb.* **2025**, *40*, 77–90. [CrossRef]
54. Eckhard, K.; Erichsen, T.; Stratmann, M.; Schuhmann, W. Frequency-dependent alternating-current scanning electrochemical microscopy (4D AC-SECM) for local visualisation of corrosion sites. *Chem. Eur. J.* **2008**, *14*, 3968–3976. [CrossRef]
55. Liu, D.; Zeng, X.; Liu, S.; Wang, S.; Kang, F.; Li, B. Application of alternating current Scanning electrochemical microscopy in lithium-ion batteries: Local visualization of the electrode surface. *ChemElectroChem* **2019**, *6*, 4854–4858. [CrossRef]
56. Gharbi, O.; Ngo, K.; Turmine, M.; Vivier, V. Local electrochemical impedance spectroscopy: A window into heterogeneous interfaces. *Curr. Opin. Electrochem.* **2020**, *20*, 1–7. [CrossRef]

https://doi.org/10.3390/coatings16020165

57. Szociński, M.; Darowicki, K.; Schaefer, K. Application of impedance imaging to evaluation of organic coating degradation at a local scale. *J. Coat. Technol. Res.* **2013**, *10*, 65–72. [CrossRef]
58. Wang, W.; Xu, L.; Sun, H.; Li, X.; Zhao, S.; Zhang, W. Spatial resolution comparison of AC-SECM with SECM and their characterization of self-healing performance of hexamethylene diisocyanate trimer microcapsule coatings. *J. Mater. Chem. A* **2015**, *3*, 5599–55607. [CrossRef]

Disclaimer/Publisher's Note: The statements, opinions and data contained in all publications are solely those of the individual author(s) and contributor(s) and not of MDPI and/or the editor(s). MDPI and/or the editor(s) disclaim responsibility for any injury to people or property resulting from any ideas, methods, instructions or products referred to in the content.

https://doi.org/10.3390/coatings16020165

coatings

MDPI

Article

Electrochemical Corrosion Performance of TiN, TiCN and TiBN Multilayer Coatings on Hardmetal Substrates

Mateja Šnajdar [1,*,†], Marin Kurtela [2,†], Danko Ćorić [2] and Matija Sakoman [2]

[1] Department of Polytechnics, Dr. Franjo Tuđman Defense and Security University (SOIS FT), Ilica 256 b, 10000 Zagreb, Croatia

[2] Faculty of Mechanical Engineering and Naval Architecture, University of Zagreb, Ivana Lučića 5, 10000 Zagreb, Croatia; marin.kurtela@fsb.hr (M.K.); danko.coric@fsb.hr (D.Ć.); msakoman@gmail.com (M.S.)

* Correspondence: mateja.snajdar@sois-ft.hr; Tel.: +385-98-415-442

† These authors contributed equally to this work.

Abstract

Three types of gradient plasma-assisted chemical vapour deposition (PACVD) coatings were produced on WC-Co hardmetal substrates: a TiN coating, a gradient TiCN coating with alternating TiN/TiCN layers and a multilayer TiBN system of TiN/TiB_2 layers. Their corrosion behaviour in a chloride environment was compared using direct current and alternating current electrochemical techniques. Potentiodynamic polarization, linear polarization and electrochemical impedance spectroscopy were carried out in 3.5 wt.% NaCl at temperature 20 ± 2 °C in a three-electrode cell with a saturated calomel electrode (SCE) reference. After 1000 s open circuit stabilization, TiN coating showed superior corrosion resistance with E_{corr} = 15 mV vs. SCE, versus TiCN (E_{corr} = −281 mV) and TiBN (E_{corr} = −304 mV). Linear polarization resistance/Tafel analysis showed significantly higher polarization resistance of TiN (R_p = 1559 kΩ·cm^2) than TiCN (195.4 kΩ·cm^2) and TiBN (243.6 kΩ·cm^2), with the lowest corrosion current density, j_{corr} = 10.97 nA·cm^{-2} and corrosion rate v_{corr} = 117.2 × 10^{-6} mm·y^{-1}. TiCN showed the highest j_{corr} (360.8 nA·cm^{-2}) and v_{corr} (3.32 × 10^{-3} mm·y^{-1}). Electrochemical impedance spectroscopy fitting with a R(QR) circuit confirmed this, with the highest charge transfer resistance at the substrate–electrolyte interface (R_{ct}) for TiN (8.198 × 10^4 Ω·cm^2), lower for TiBN (7.929 × 10^4 Ω·cm^2) and lowest for TiCN (1.435 × 10^4 Ω·cm^2), indicating TiN as the best barrier and TiCN as the most permeable.

Keywords: PACVD; corrosion; coating; potentiodynamic polarization; linear polarization (LPR); electrochemical impedance spectroscopy (EIS)

1. Introduction

Tungsten carbide–cobalt (WC-Co) hardmetals are the most widely used type of hardmetals, intended for applications such as cutting and mining tools, forming and similar components where high hardness, mechanical strength and wear resistance are needed [1–3]. Their superior properties result from very hard WC grains embedded in a tougher Co binder phase which ensures toughness [4–6]. Despite their superior mechanical properties, WC-Co hardmetals have been shown to be susceptible to corrosion, especially in chloride-containing environments, where selective dissolution of Co binder may occur [7–11]. Such corrosion-induced degradation can significantly reduce service life by accelerating wear, impairing parts' integrity and promoting premature failure.

To address this, different surface modification techniques for hard coating deposition have become a dominant strategy of extending WC-Co components' functional lifetime [12–14]. Among the available coating technologies, plasma-assisted chemical vapour deposition (PACVD) has attracted increased interest, predominantly due to its ability to produce dense, adherent coatings at relatively low temperatures [15–19]. This feature is especially important for coating nanostructured hardmetal substrates, as high-temperature technologies could induce grain coarsening, η-phase formation and decarburization, all of which lead to the deterioration of mechanical properties and chemical stability [20–22]. PACVD technology enables the deposition of hard nitride, carbonitride and boride-based coatings while preserving the desirable microstructural properties of the substrate [15,16,23,24].

Titanium-based coatings such as titanium nitride (TiN), titanium carbonitride (TiCN) and titanium boron nitride (TiBN) are among the most applied PACVD coatings on WC-Co hardmetal substrates. Their influence on hardness, wear resistance, adhesion and general tribological performance has been extensively studied [25–27]. Complex multilayer and gradient coating architectures have been shown to significantly improve adhesion, reduce residual stresses and limit crack propagation. Previous studies have demonstrated superior properties of complex TiCN and TiBN coatings in demanding wear and cutting applications compared to TiN coatings under the same conditions [18,28,29].

While the mechanical and tribological behaviour of PACVD-coated WC-Co substrates has been widely reported, their corrosion behaviour has yet to be fully investigated. The WC-Co components' corrosion resistance is of great importance, since they are often exposed to aggressive environments during service, including humid atmospheres, saline solutions and lubricants. During operation, corrosion and wear often act simultaneously, thus leading to accelerated material degradation [30,31]. Therefore, understanding the corrosion mechanisms of coated WC-Co systems is essential for the reliable design of surface-modified components.

The corrosion behaviour of coated WC-Co materials is not only governed by the chemical stability of the coating itself, but also by its thickness, adhesion to the substrate material, microstructural characteristics, and the presence of localized defects. These defects, in the form of microcracks, pores or weak interfaces between layers, can become pathways for aggressive media penetration, therefore leading to corrosion at the substrate/coating interface [32,33]. In the case of complex multilayer coatings, internal interfaces between individual layers can additionally influence corrosion due to potential lateral diffusion of corrosive media [30].

Recent studies have already described the mechanical, physical and tribological behaviour of WC-Co systems that contain TiN, TiCN, and TiBN layers, the results of which suggest a strong influence of coating architecture on coatings' mechanical integrity and adhesion [15–18]. However, the corrosion behaviour of these different types of coating systems has not yet been systematically investigated. This study addresses this gap by evaluating the corrosion behaviour using standardized electrochemical methods. The electrochemical results are interpreted in relation to the already available data on structural and mechanical properties, allowing correlations between coating integrity, defect influence and resulting electrochemical response. The conducted research is aimed at clarifying how coating architecture and properties influence corrosion mechanisms in the case of multilayer gradient PACVD-coated WC-Co hardmetals. By comparing coatings under the same electrochemical conditions, a new insight into the role of coating design on corrosion protection is provided, in particular monolithic vs. multilayer coating design. The conclusions of this study will contribute to our understanding of surface-engineered WC-Co materials and support further development of corrosion-resistant coating systems

https://doi.org/10.3390/coatings16030353

with a prolonged lifetime. In addition, the results obtained in this study make a significant contribution to future comparisons of how different coating deposition techniques influence the electrochemical and corrosion performance of protective coatings.

2. Materials and Methods

2.1. Sample Consolidation and Coating

For the present study, nanostructured WC-Co hardmetals with 10 wt.% Co were examined. Tungsten carbide (mean particle size 0.095 μm and specific surface area 3.92 m^2/g, H.C. Starck, Goslar, Germany) was mixed with cobalt powder (mean particle size 0.640 μm, specific surface area 2.96 m^2/g, Umicore, Markham, ON, Canada). Vanadium carbide (0.5 wt.% VC) and chromium carbide (0.75 wt.% Cr_2C_3) were used as grain-growth inhibitors with carbon balance tailored to suppress the formation of free carbon and η-phase during sintering. Powders were dry mixed in a horizontal ball mill with the addition of paraffin. Sample consolidation was conducted via the sinter-HIP technique in an FCT Anlagenbau furnace (type FPW 280/600-3-2200-100-PS, Sonnenberg, Germany). Consolidation comprising a debining/pre-sintering step was conducted under a controlled H_2/Ar atmosphere, followed by two-step densification (vacuum sintering at 1350 °C and hot isostatic pressing at the same temperature under an argon pressure of 100 bar [15]. Earlier investigations confirmed that the described sintering route resulted in fully dense nanostructured substrates without unbound carbon or η-phase, as verified by magnetic saturation and coercivity measurements, microstructural analysis, and XRD [15–18].

All samples were prepared before coating using standard procedures for WC-Co substrates (grinding and polishing) to ensure comparable surface conditions across different sample groups.

Coating was conducted via a PACVD process at 530 °C using the Rübig PC 70/90 system (Rübig GmbH and Co KG, Marchtrenk, Austria). The coatings studied in this paper are the following architectures: a monolithic TiN coating (TiN sample), a multilayer TiN/TiCN coating (TiCN sample) and a gradient multilayer TiN/TiB_2/TiBN coating (TiBN sample) [15–18]. The measured TiN coating thickness was of ~3 μm, while the coating thicknesses were ~5 μm and ~1 μm for TiCN and TiBN, respectively. Surface roughness was measured for all samples and showed submicron arithmetic mean roughness values, which were important for discarding the influence of macroscopic roughness effects on the corrosion behaviour studied here. Detailed mechanical characterization of coated samples including instrumented nanoindentation, microhardness testing, Rockwell adhesion testing and scratch testing was previously conducted, with results reported [15]. Quantitative depth profile (QDP) analysis was performed to determine elemental distribution along the coating cross-section anngs was analyzed by XRD using an XRD6000 diffractometer (Shimadzu, Tokyo, Japan) with Cu Kα radiation (λ = 1.5406 Å). Phase identification was performed using standard reference databases. Coating thickness was measured by the Calotest method in accordance with EN 1071-2:2002 [34] using a TRIBOtechnik Calotester (TRIBOtechnik GmbH, Bad Düben, Germany) with 25 mm steel ball, under 500 rpm for 45 s [15].

TiN coating exhibited lower hardness and elastic modulus compared to TiCN and TiBN coating and inferior adhesion behaviour (HF-5 in Rockwell indentation tests), followed by low critical scratch loads [15,18]. The obtained characteristics of the TiN coating indicate limited resistance to residual and external stresses, which could promote microcrack formation and localized coating failures when exposed to corrosive media.

The results of the previous characterization of TiCN coating show higher hardness and elastic modulus along with superior adhesion compared to TiN coating. The complex multilayer architecture of the numerous TiN/TiCN interfaces was identified as a potential source of defect propagation channels, which may facilitate the transport of the electrolyte in a lateral or transversal direction [15,18].

The TiBN coating gradient multilayer architecture with a gradual transition from the TiN to the TiB_2/TiBN layers has been confirmed. This type of coating showed the highest adhesion quality and scratch resistance, indicating enhanced coating/substrate bonding and improved resistance to crack initiation and propagation, which should be crucial for preserving coating integrity under corrosion conditions.

These previously determined and reported structural and mechanical characteristics (summarized in Tables 1 and 2) provide the necessary background for interpreting the electrochemical corrosion behaviour presented in this study [15,18].

Table 1. Structural and mechanical sample properties of samples used for electrochemical testing [15,18].

Sample	Coating Architecture	Thickness [μm]	Surface Roughness, R_a [μm]	Microhardness HV 0.005	Elastic Modulus, E_{IT} [GPa]	Adhesion Quality (Rockwell HF)	Scratch Test Critical Load L_{c2} [N] [1]
WC-10 wt.% Co-TiN	single-layer TiN	3.10 ± 0.23	0.150 ± 0.009	2184 ± 61	336 ± 35	HF-5	~27.11 ± 5.30
WC-10 wt.% Co-TiCN	multilayer TiN/TiCN (TiCN top)	5.33 ± 0.25	0.154 ± 0.021	3220 ± 36	398 ± 15	HF-3	35.52 ± 7.18
WC-10 wt.% Co-TiBN	gradient multilayer TiN/TiB_2/TiBN	1.63 ± 0.23	0.146 ± 0.003	3672 ± 135	466 ± 25	HF-1	no delamination observed

[1] L_{c2}—critical delamination load [N] associated with the onset of adhesive coating failure.

Table 2. Coating architecture and composition analysis results [15].

Coating Type	QDP Analysis Characteristics	XRD Detected Phases	Layer Function
TiN	- Ti and N dominant up to 3 μm where sharp transition to W, C, Co occurs - no gradient region	- cubic TiN - WC substrate peaks (hexagonal)	- adhesion layer and protective coating - residual stress reduction
TiCN	- Ti, N, C present up to 5.2 μm - higher N near substrate (TiN base layer) - higher C near surface (TiCN top layer) - gradient N → C transition	- TiN (cubic) - TiC/TiC0.7N0.3 (cubic) - WC substrate with possible overlapping TiCN	- TiN: adhesion and stress absorber - gradient TiN/TiCN: stress relaxation - top TiCN: hardness and wear resistance
TiBN	- B content highest at surface (TiB_2 top layer) - N peak detected at ~0.3 μm (TiN base layer) - gradual TiN → TiB_2 transition - interface reached at ~2 μm	- TiB_2 (hexagonal) - TiN (cubic)	- TiN: adhesion and thermal - expansion match - gradient TiN/TiB_2: improved cohesion - top TiB_2: highest hardness and modulus

2.2. Electrochemical Characterization

The corrosion resistance of samples TiN (WC-10 wt.% Co + TiN coating), TiCN (WC-10 wt.% Co + TiCN coating), and TiBN (WC-10 wt.% Co + TiBN coating) was evaluated using both direct current (DC) and alternating current (AC) electrochemical techniques. DC methods provide the data of corrosion potential, corrosion current density and polarization resistance, which are used to describe corrosion kinetics [35,36]. For the characterization

https://doi.org/10.3390/coatings16030353

of barrier properties and interfacial processes, AC methods are used, electrochemical impedance spectroscopy (EIS) being the most common [37]. By combining both AC and DC methods, a comprehensive insight into the corrosion performance of the coatings and present corrosion mechanisms is obtained [7,38,39]. Potentiodynamic polarization and EIS measurements were performed using a VersaSTAT 3 potentiostat/galvanostat (AMETEK Scientific Instruments, Princeton Applied Research, Berwyn, PA, USA) controlled by VersaStudio software (version 2.6; Princeton Applied Research, AMETEK Scientific Instruments, Oak Ridge, TN, USA). All experiments were conducted in accordance with EN ISO 17475:2019 [40]. The measurements were performed in a conventional three-electrode electrochemical cell, consisting of the sample as the working electrode, an Ag/AgCl or saturated calomel electrode (SCE) as the reference electrode, and a graphite rod as the auxiliary electrode. A 3.5 wt.% NaCl aqueous solution was used as the electrolyte, and the room temperature (ϑ_s) was maintained at 20 $\pm$ 2 °C. Representative micrographs of the surface condition before and after electrochemical testing are provided for the sample exhibiting the highest corrosion rate. Surface analysis was performed using an optical microscope (GX51F, Olympus, Tokyo, Japan) and scanning electron microscope (VEGA, TESCAN, Brno, Czech Republic).

3. Results

3.1. Results of Electrochemical DC Measurements

At the beginning of the polarization measurements, the electrochemical system was allowed to stabilize after immersion of the samples in a 3.5 wt.% NaCl aqueous solution, with the electrical circuit between the working and counter electrodes kept open.

The potential difference between the working and reference electrodes was continuously recorded as a function of time. The open circuit potential (E_{corr}) was determined after a stabilization period of 1000 s. In addition to the open circuit potential measurements, the polarization resistance (R_p) and corrosion rate (v_{corr}) were evaluated using linear polarization resistance and Tafel polarization techniques. Anodic and cathodic polarization curves were recorded over a potential range of ($E = E_{corr} \pm 0.25$ V). The corrosion current density j_{corr} was determined by extrapolating the linear regions of the anodic and cathodic Tafel plots. All measurements were carried out at a scan rate of 0.167 mV·s^{-1}, and the polarization data are presented graphically in logarithmic form ($E - log$ j). The corrosion rate was calculated from corrosion current density values using the equivalent weight (EW) and density ρ (g·cm^{-3}) of the material. The obtained numerical values of open circuit potentials, linear polarization resistance and corrosion rate of uncoated substrate [7] and coated samples are summarized in Table 3, while the corresponding graphical representations are shown in Figures 1–3.

Table 3. The results of electrochemical DC techniques.

Sample	ϑ_s [°C]	E_{corr} vs. SCE [mV]	R_p [kΩ·cm^2]	β_a * [mV/dec]	β_c ** [mV/dec]	j_{corr} [nA·cm^{-2}]	v_{corr} [mm·y^{-1}]
substrate WC-Co	20 $\pm$ 2	−308	452.8	98.34	97.7	36.6 $\times$ 10^3	388.8 $\times$ 10^{-3}
TiN	20 $\pm$ 2	15	1559	237	119.5	10.97	117.2 $\times$ 10^{-6}
TiCN	20 $\pm$ 2	−281	195.4	223	201	360.8	3.32 $\times$ 10^{-3}
TiBN	20 $\pm$ 2	−304	243.6	60.4	201.9	21.43	322.8 $\times$ 10^{-6}

* β_a—anodic Tafel slope, describes anodic reaction kinetics. ** β_c—cathodic Tafel slope, describes cathodic reaction kinetics.

https://doi.org/10.3390/coatings16030353

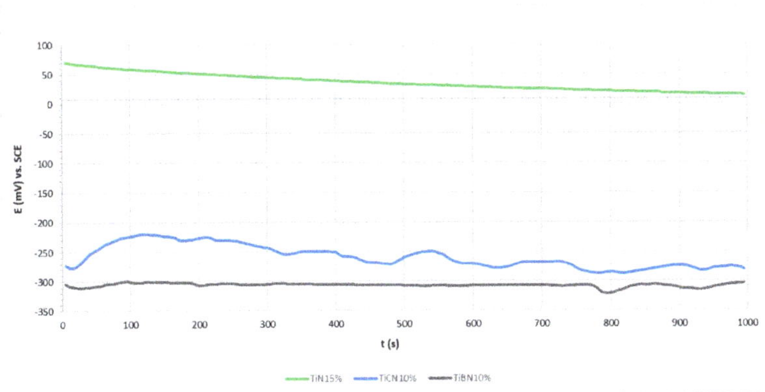

Figure 1. Open circuit potential versus time curves.

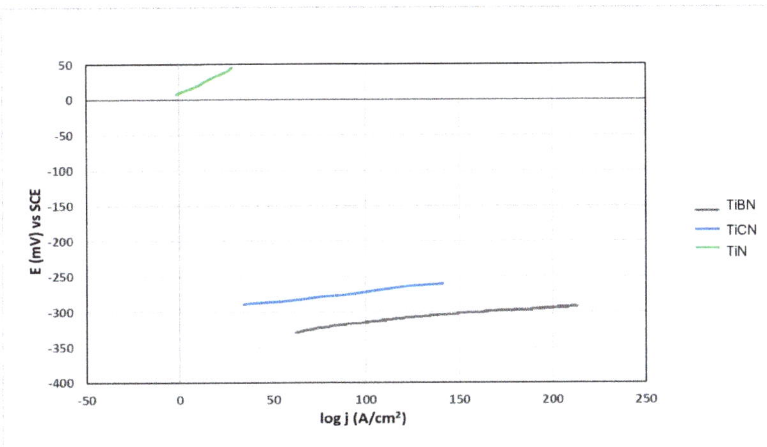

Figure 2. Linear polarization curves.

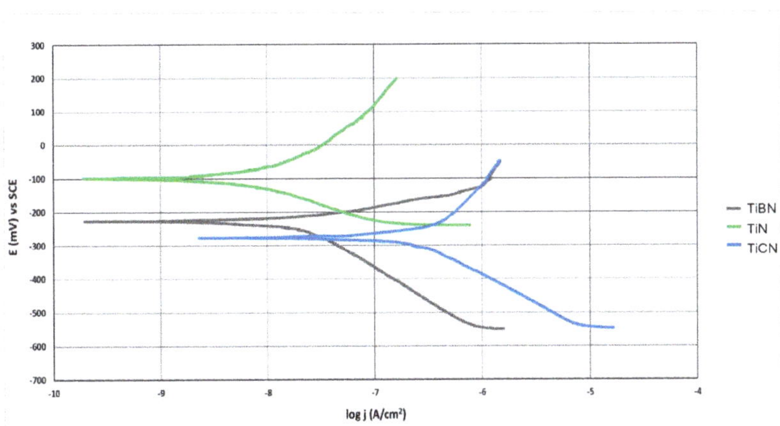

Figure 3. Potentiodynamic polarization curves (Tafel extrapolation diagram).

https://doi.org/10.3390/coatings16030353

The linear polarization resistance curves, which characterize the electrochemical response of the samples in the vicinity of the open circuit potential ($E = E_{corr} \pm 20$ mV), are shown in Figure 2.

The potentiodynamic polarization curves, presented as Tafel extrapolation plots, used to determine the corrosion rate are shown in Figure 3.

3.2. Results of Electrochemical AC Measurements

The EIS method was employed to evaluate the corrosion resistance and protective performance of the investigated samples. The use of an AC technique enabled the determination of surface layer resistance and the identification of corrosion mechanisms without inducing damage to the tested surfaces. To facilitate faster and more practical data interpretation, several simplified EIS spectrum analysis methods reported in the literature were also applied. The resistance of both the substrate and the protective barrier layers is predominantly reflected in variations in resistance and capacitance parameters. Accordingly, degradation of the surface layers was inferred from changes in these parameters, which manifested as variations in the overall impedance response. To accurately describe the electrochemical reactions occurring at the interface between the working electrode and the electrolyte, an appropriate equivalent electrical circuit composed of resistive and capacitive elements was selected. This modelling approach enables a precise interpretation of the corrosion mechanism. A metal–electrolyte interface undergoing simple oxidation or reduction reactions can be appropriately represented by the equivalent circuit R(QR), as shown in Figure 4. The circuit consists of the solution (ohmic) resistance of the electrolyte (R_s), the charge transfer (polarization) resistance at the substrate–electrolyte interface (R_{ct}), and a constant phase element (CPE) characterized by the parameter Q.

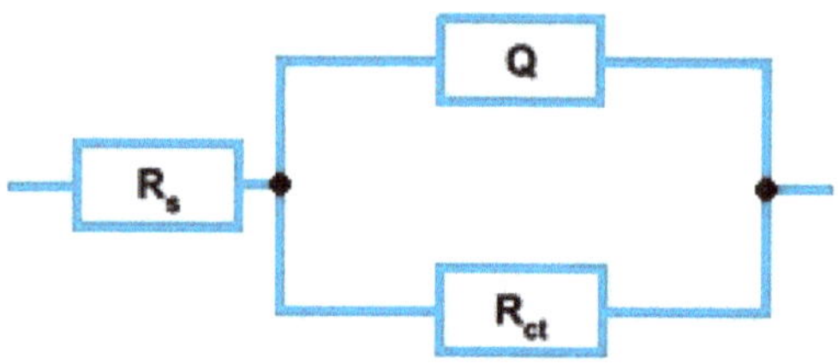

Figure 4. Model of R(QR) equivalent electrical circuit. Reprinted from Ref. [41].

The constant phase element is incorporated into the equivalent electrical circuit to compensate for system inhomogeneities associated with surface roughness, inhibitor adsorption, and the porosity of the newly formed film. The impedance of the CPE (Z_{CPE}) is described by the following equations [42]:

$$Z_{CPE} = Q^{-1}(j\omega)^{-n}, \tag{1}$$

$$C = Q(\omega_{max})^{n-1}, \tag{2}$$

in which Q denotes the proportionality constant (CPE), n the empirical exponent or phase shift ($-1 \leq n \leq 1$), ω the angular frequency, j the root of -1 (imaginary unit), C the capacity, and ω_{max} the angular frequency at which the imaginary component impedance Z'' reaches the maximum of the time constant. In this way, the impedance spectrum of a distributed system is explained, which cannot be interpreted by a finite number of ideal electrical elements. In the case when $0.8 < n \leq 1$, CPE represents a capacitor of capacity Q, for $n = 1$, CPE represents an ideal coil $Q = L$, and for $n = 0.5$, CPE represents Warburg impedance $Q = W$, while for values of $n = 0$, CPE represents an ideal resistor $Q = R$.

https://doi.org/10.3390/coatings16030353

The measurements were performed at room temperature (20 ± 2) °C using a 3.5 wt.% NaCl aqueous solution as the electrolyte. Impedance spectra were recorded at the open circuit potential with a sinusoidal voltage amplitude of 10 mV over a frequency range of 100 kHz to 10 mHz. Data fitting, i.e., matching the measured responses to the elements of the equivalent circuit, was carried out using ZSimpWin software (Version 3.2).

The Nyquist and Bode plots of the tested samples, along with the corresponding equivalent circuit model employed to simulate the EIS results, are presented in Figures 5 and 6. These diagrams provide a comprehensive representation of the impedance behaviour of the samples over the applied frequency range. The Nyquist plots illustrate the relationship between the real and imaginary components of the impedance, highlighting the capacitive and resistive characteristics of both the substrate and the protective surface layers.

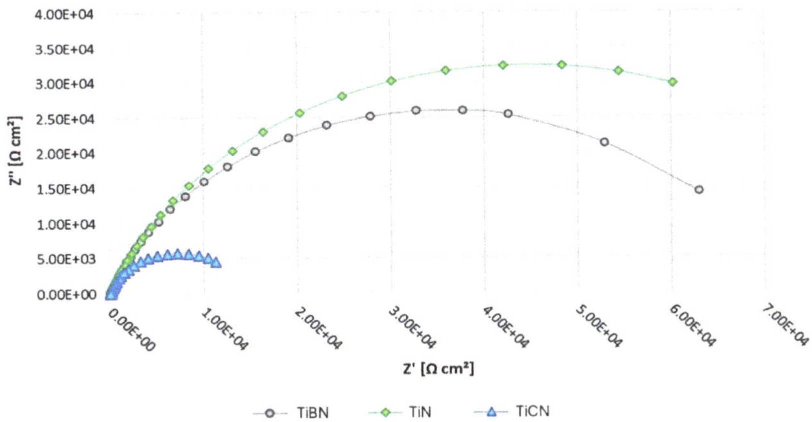

Figure 5. Nyquist plot of coated samples.

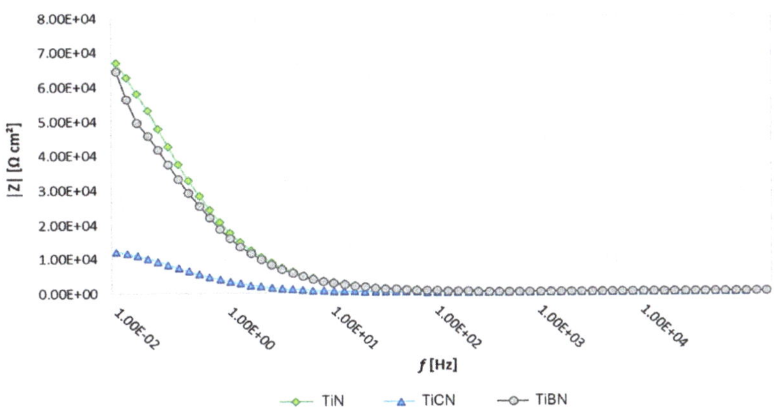

Figure 6. Bode plot of coated samples.

The Bode plots, on the other hand, show the frequency-dependent behaviour of the impedance magnitude and phase angle, which allows for the identification of processes occurring at different time scales, such as charge transfer, diffusion, and barrier layer effects. Figure 7 further details the EIS response across high, medium, and low frequencies, emphasizing how different frequency regions correspond to distinct electrochemical phenomena. High-frequency responses are generally dominated by solution resistance and surface roughness effects and medium frequencies primarily reflect charge transfer

https://doi.org/10.3390/coatings16030353

and interfacial reactions, while low-frequency behaviour provides insight into diffusion-controlled processes and the integrity of the protective layers. Together, these figures offer a complete overview of the electrochemical performance of the samples and validate the applicability of the selected equivalent circuit model for accurately interpreting the corrosion mechanisms.

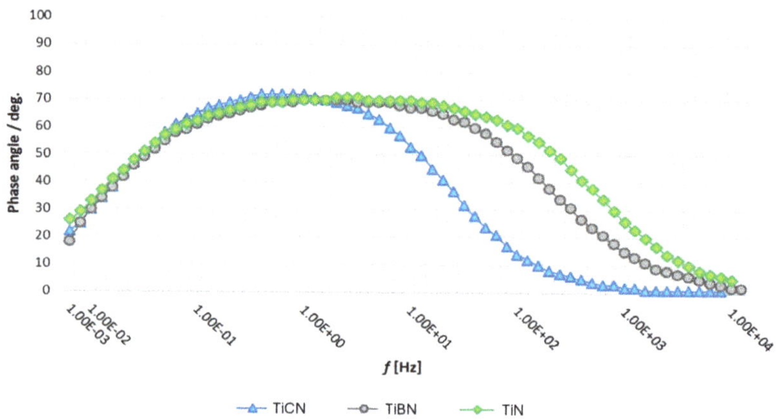

Figure 7. Phase angle as a function of frequency.

In electrochemical impedance spectroscopy (EIS), the chi-square (χ^2) factor is a statistical parameter that quantifies the goodness of fit between the experimental data and the equivalent electrical circuit (EEC) model. $\chi^2 \leq 10^{-3}$ in this work indicates that the selected equivalent circuit properly represents the electrochemical behaviour of the coating system. The results of the electrochemical impedance spectroscopy measurements of coated samples and substrate WC-Co material [7] are presented in Table 4.

Table 4. The results of the EIS technique.

Sample	ϑ_s [°C]	R_s [Ω cm^2]	Q [F cm^2]	n_1	R_{ct} [Ω cm^2]
substrate WC-Co	20 ± 2	4.504	2.213×10^{-3}	0.73	8.068×10^{-2}
TiN	20 ± 2	3.978×10^{3}	9.173×10^{-5}	0.80	8.918×10^{4}
TiCN	20 ± 2	3.978×10^{3}	5.704×10^{-5}	0.86	1.435×10^{4}
TiBN	20 ± 2	3.978×10^{3}	1.956×10^{-5}	0.79	7.929×10^{4}

Figure 8 shows optical and SEM micrographs obtained for the sample with the highest measured corrosion rate during the electrochemical corrosion tests, the TiCN-coated WC-Co sample. No significant morphological changes, such as surface degradation or the accumulation of corrosion products, are observed. The electrochemical tests performed in this study are short-term experiments performed in 3.5 wc. % NaCl solution. The corrosion current densities obtained for the coated samples are in the nanoampere range, indicating very low corrosion rates of ~10^{-4} to 10^{-6} mm/year. Under such low corrosion rates and short exposure time, the total amount of dissolved material is extremely low, so visible corrosion damage or corrosion products detectable by SEM or optical microscope imaging was not expected.

https://doi.org/10.3390/coatings16030353

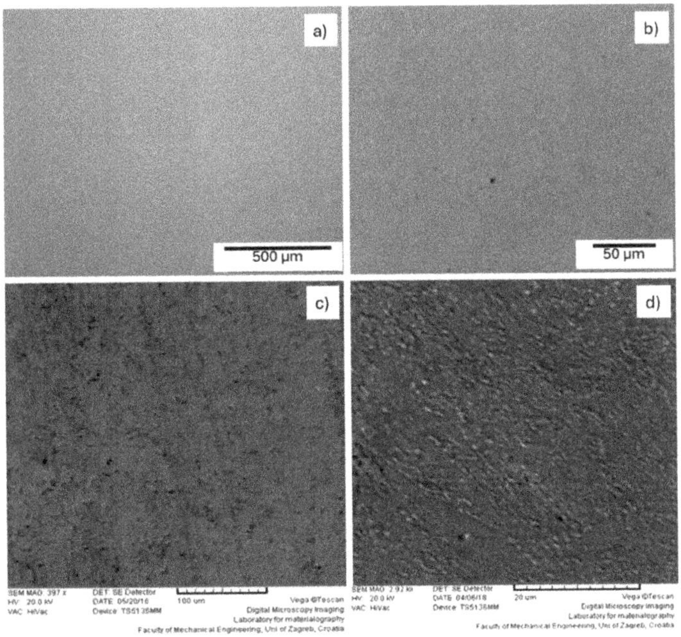

Figure 8. Optical and SEM micrographs of sample WC-10 wt.% Co + TiCN coating: (**a**) optical micrograph before electrochemical testing; (**b**) optical micrograph after electrochemical testing; (**c**) SEM micrograph before testing; (**d**) SEM micrograph after testing.

4. Discussion

From the measured numerical values of the open circuit potential shown in Table 3 and graphically interpreted in Figure 1, information on the corrosion behaviour of the working electrodes in the test electrolyte was obtained. After an approximately established stationary state (t = 1000 s), when the test samples are in equilibrium with the environment (anodic and cathodic currents are equal in magnitude, but in opposite directions), it is visible that sample TiN (15 mV vs. SCE) has a more positive E_{corr} than sample TiCN (−281 mV vs. SCE) and TiBN (−304 mV vs. SCE), i.e., it is more resistant to corrosion for the same test solution and temperature. It can also be seen that, for measurements in 3.5 wt.% NaCl aqueous solution, the potentials of sample TiN to samples TiCN and TiBN slightly shift in the negative direction. Such potential oscillation can be correlated with corrosion activity (instability on the surfaces and material dissolution) of the samples. All measurements exhibited excellent repeatability, confirming the reliability of the experimental setup and the stability of the electrochemical system during testing.

Polarization resistance measurements offer a rapid and reliable method for evaluating the corrosion behaviour of materials, typically requiring less than ten minutes to complete. This technique is particularly advantageous because it allows for the direct estimation of corrosion rates and has been shown to correlate well with conventional long-term methods, such as weight loss determinations, providing a practical alternative for comparative studies of different substrates. In the present study, the polarization resistance was determined by analyzing the slope of the linear region of the anodic and cathodic branches of the polarization curves, as illustrated in Figure 2. This slope corresponds to the resistance to charge transfer at the electrode–electrolyte interface, which is inversely proportional to the corrosion rate according to the Stern–Geary relationship.

https://doi.org/10.3390/coatings16030353

The results indicate that sample TiN (1559 kΩ cm^2) exhibits a significantly higher polarization resistance compared to samples TiCN (195.4 kΩ cm^2) and TiBN (243.6 kΩ cm^2). Quantitatively, the R_p of sample TiN is approximately seven times greater than that of the other two samples, reflecting a substantially enhanced ability of the surface to resist electrochemical attack. This higher resistance can be attributed to the formation of a more stable and uniform barrier layer on the surface of the TiN sample, which effectively limits both anodic dissolution and cathodic reactions. In contrast, the lower R_p values observed for samples TiCN and TiBN suggest less effective surface coverage, which facilitates the penetration of the electrolyte and accelerates corrosion processes.

Furthermore, the polarization curves provide additional insight into the electrochemical behaviour of the surface layer. The linearity and slope of the curves in the vicinity of the open circuit potential reflect the efficiency of the protective layer in impeding electron transfer. In the TiN sample, the broader and more distinct linear region indicates a more stable passive behaviour, while the steeper slopes for samples TiCN and TiBN correspond to faster corrosion kinetics. These findings are consistent with the observed trends in open circuit potential measurements and confirm the superior corrosion protection under the tested conditions. Polarization resistance measurements demonstrate both the quantitative and qualitative differences in the corrosion performance, highlighting the critical influence of composition and structure on the electrochemical stability of the samples. From the numerical values of the corrosion current density j_{corr} (Table 3), obtained from the Tafel plot by extrapolating the linear portion of the curve to E_{corr}, as shown in Figure 3, it can be concluded that sample TiN (117.2 × 10^{-6} mm/y), compared to samples TiCN (3.32 × 10^{-3} mm/y) and TiBN (322.8 × 10^{-6} mm/y), has a lower corrosion rate, which unambiguously indicates a higher resistance of the surface layers, caused by the chemical composition and surface condition. Also, sample TiN has the lowest corrosion current density of 10.97 nA/cm^2 and sample TiBN has the average corrosion current density value of 21.43 nA/cm^2, while the highest corrosion current density of 360.8 nA/cm^2 was measured for sample TiCN with a different microstructure and grain size.

From the numerical values of the corrosion current density, j_{corr} (Table 3), obtained from Tafel plots by extrapolating the linear portion of the curves to the corrosion potential, E_{corr} (Figure 3), it is evident that the corrosion behaviour of the investigated samples differs significantly. The uncoated WC-Co substrate shows active corrosion behaviour, with a corrosion rate of 388.8 × 10^{-3} mm/y, whereas the coated samples show a substantial improvement in corrosion resistance. The TiN sample exhibits a corrosion rate of 117.2 × 10^{-6} mm/y, which is moderately lower than that of sample TiBN (322.8 × 10^{-6} mm/y) and significantly lower than that of sample TiCN (3.32 × 10^{-3} mm/y). This indicates that the surface layer of the TiN sample provides the highest corrosion resistance, likely due to its specific chemical composition, surface structure, and favourable microstructural features that inhibit the electrochemical degradation process.

When examining the corrosion current densities, further insights into the electrochemical stability are obtained. The TiN sample exhibits the lowest corrosion current density of 10.97 nA/cm^2, indicating its exceptional ability to resist charge transfer reactions that lead to corrosion. The TiBN sample shows an intermediate corrosion current density of 21.43 nA/cm^2, suggesting moderate resistance, whereas TiCN exhibits a significantly higher corrosion current density of 360.8 nA/cm^2. The high current density in the TiCN sample is consistent with its higher corrosion rate and can be attributed to differences in microstructure, grain size, and potentially less dense or more defective surface layers that facilitate the ingress of corrosive species.

These results highlight that the corrosion resistance of the surface is strongly affected not only by its chemical composition but also by microstructural characteristics and surface conditions. The combination of low corrosion current density and low corrosion rate in the TiN sample demonstrates that the sample surface can provide effective protective performance, making it suitable for applications requiring enhanced durability in corrosive environments. Comparatively, the higher corrosion activity observed in samples TiCN and TiBN underscores the importance of optimizing both composition and microstructure to achieve improved surface protection.

Based on the electrochemical impedance spectroscopy results, the corrosion resistance of samples TiN, TiCN and TiBN samples in 3.5 wt.% NaCl aqueous solution can be quantitatively evaluated and unambiguously compared. EIS provides insight into the electrochemical processes occurring at the interface between the substrate and the electrolyte by representing the system with an equivalent electric circuit. This circuit typically consists of resistive and capacitive elements that model charge transfer, double-layer capacitance, and other interfacial phenomena. By fitting the measured impedance data to this equivalent circuit, it is possible to extract numerical values that reflect the electrochemical behaviour and stability of each sample, allowing for a detailed interpretation of the corrosion mechanisms.

The results, summarized in Table 4, show a clear correlation with the previously obtained DC measurements, reinforcing the statement that TiN exhibits the highest corrosion resistance. This is confirmed by the significantly higher resistance to charge transfer, $R_{ct} = 8.918 \times 10^4\ \Omega\ cm^2$ at the substrate–electrolyte interface. In comparison, TiBN shows a moderately lower $R_{ct} = 7.929 \times 10^4\ \Omega\ cm^2$, while TiCN has the lowest value of $R_{ct} = 1.435 \times 10^4\ \Omega\ cm^2$, reflecting a much higher susceptibility to corrosion. These differences highlight the influence of both chemical composition and surface condition on the electrochemical stability of the samples. When comparing the results of the coated samples with those obtained for the substrate WC-Co sample [7], a significant improvement in corrosion resistance was achieved by coating regardless of the coating type. The uncoated substrate exhibits very low charge transfer resistance ($R_{ct} = 8.068 \times 10^{-2}\ \Omega\ cm^2$), indicating strong electrochemical activity and poor corrosion resistance. In contrast, all coated samples show that R_{ct} values increase by several orders of magnitude, confirming a significant reduction in corrosion rate.

The Nyquist diagrams presented in Figure 5 illustrate these observations. Each sample shows a capacitive semicircle whose radius corresponds to the charge transfer resistance. The TiN sample exhibits the largest semicircle diameter, indicating the most efficient barrier to electron transfer and, therefore, the highest corrosion resistance. The TiBN sample presents a smaller semicircle, suggesting intermediate protection, whereas TiCN exhibits the smallest semicircle, consistent with its significantly lower resistance to charge transfer. The progressive decrease in the diameter of the semicircle from samples TiN and TiBN to TiCN demonstrates a direct relationship between the size of the capacitive loop and the effectiveness of the protective surface layer.

Overall, the EIS analysis confirms that TiN provides superior protection in chloride-containing environments due to a combination of high interfacial resistance and stable surface characteristics. TiBN offers moderate protection, while TiCN, with its comparatively low R_{ct} and small semicircle radius, is the least resistant to corrosion. These results emphasize the importance of optimizing both the material composition and microstructural properties to enhance the long-term performance of the surface in aggressive aqueous environments. The detailed EIS interpretation also provides a deeper understanding of the electrochemical interactions occurring at the phase boundary, which is critical for predicting durability and designing more corrosion-resistant materials.

The electrochemical impedance data further illustrate the superior corrosion resistance of the TiN coating. In particular, the higher charge transfer resistance, $R_{ct} = 8.918 \times 10^4\ \Omega\ cm^2$, combined with the lower double-layer capacitance, $Q = 9173 \times 10^{-5}$ F cm^2, is a clear

https://doi.org/10.3390/coatings16030353

indication of the limited penetration of the electrolyte into the surface layer. A high R_{ct} implies that the electron transfer reactions necessary for corrosion are significantly hindered, while a low capacitance value indicates that the surface layer is dense and uniform, minimizing localized electrolyte accumulation at the interface.

The Bode plot analysis provides additional confirmation of these observations. In the frequency-dependent impedance representation, TiN exhibits the highest impedance modulus at low frequencies and a relatively low phase angle deviation, suggesting minimal degradation of the surface layer over the measured frequency range. Specifically, in Figure 6, the plateau observed at low frequency (10 mHz) for the TiN sample is elevated compared to the TiCN and TiBN samples, which directly correlates with its higher R_{ct} and lower Q values. The slope of the Bode curves for the TiN and TiBN samples remains nearly constant across the intermediate frequency range, indicating stable capacitive behaviour and consistent electrochemical performance of the surface layers. In contrast, TiCN shows lower R_{ct} values and higher capacitances, corresponding to more pronounced electrolyte absorption and less effective surface protection. The elevated capacitance indicates increased interfacial charge storage, which is typically associated with a partially degraded surface layer that allows the electrolyte to penetrate more easily, facilitating corrosion processes. Consequently, this sample demonstrates lower overall impedance and a reduced ability to resist electrochemical degradation in chloride-containing environments. Overall, the combination of high charge transfer resistance, low capacitance, and the characteristic Bode plot profile of sample TiN confirms that it maintains the structural and electrochemical integrity of the surface layer most effectively. These findings highlight the importance of both surface composition and microstructural uniformity in achieving long-lasting corrosion resistance and emphasize that EIS, particularly the analysis of R_{ct}, Q, and frequency-dependent impedance behaviour, is a powerful tool for quantitatively assessing the protective performance of the surface layer in aggressive aqueous environments.

The phase angle analysis and its corresponding frequencies, presented in Figure 7, provide additional insight into the electrochemical behaviour and protective performance of the coated samples. Phase angles close to 90° are characteristic of ideal capacitors. All samples exhibit relatively high phase angles, approaching 75°, which demonstrates a predominantly capacitive response and suggests the presence of well-formed surface layers that act as barriers to charge transfer. A high phase angle over a broad frequency range is typically associated with a stable and dense surface layer that effectively hinders electrolyte penetration, reinforcing the observations obtained from the Nyquist and Bode plots. A closer examination of the phase angle maxima reveals differences in the frequency at which these peaks occur. For samples TiN and TiBN, the maximum phase angles are observed at medium frequencies around 5 Hz, whereas for sample TiCN, the maximum occurs at lower frequencies. This indicates that the dominant time constant of the system is in the mid-frequency region, which is generally associated with the coating layer itself. Such behaviour suggests a compact coating structure, a relatively low electrolyte permeability and a stable coating/electrolyte interface, and is directly correlated with an increase in capacitance and pore resistance, reflecting a greater ability of the surface layer to store charge, indicating good protective performance. In practical terms, a lower frequency peak suggests that the surface layer is less resistive and more prone to electrolyte infiltration, which aligns with the lower charge transfer resistance observed for this sample in the EIS measurements. For TiN coating, the medium-frequency maximum, combined with a high phase angle, indicates an optimal balance between capacitive behaviour and resistance to charge transfer. This confirms that the TiN layer provides a dense and protective barrier at the electrolyte–substrate interface. The TiBN coating, while also showing a peak at medium frequencies, has slightly lower impedance values than TiN, suggesting moderate protection.

https://doi.org/10.3390/coatings16030353

In contrast, the shift in the phase angle peak to lower frequencies for TiCN indicates a less compact surface layer with increased electrolyte absorption, consistent with its higher capacitance and lower R_{ct}. Overall, the phase angle analysis complements the Nyquist and Bode plot data by providing frequency-dependent information about the capacitive behaviour of the surface layers.

The location and magnitude of the phase angle peaks serve as indicators of surface layer integrity, electrolyte penetration, and charge transfer characteristics, confirming that the TiN layer exhibits the highest electrochemical stability, followed by TiBN, with TiCN being the least resistant to corrosion in 3.5 wt.% NaCl solution.

Despite the higher hardness and elastic modulus of TiCN and TiBN coating compared to single-layer TiN coating [15], corrosion resistance is not governed solely by mechanical properties. It is predominantly influenced by coating compactness, chemical composition, defects density and electrochemical stability. The multilayer structure of TiN/TiCN and TiN/TiB_2/TiBN coating contains numerous interfaces, which may act as pathways for electrolyte penetration, especially in the case of long exposure to the corrosive medium. Also, higher stiffness, i.e., the elastic modulus of the multilayer TiCN and TiBN coatings, may be connected to the development of residual stresses during deposition and cooling, arising from differences in the thermal expansion of individual TiN/TiCN and TiN/TiB_2/TiBN layers. Such stresses can lead to the formation of microdefects that are not necessarily detectable by standard mechanical characterization but do influence electrochemical behaviour. Also, the presence of carbon in the TiCN could modify local electrochemical response and increase interfacial capacitance, as indicated by higher Q values and lower R_{ct} values obtained from EIS analysis.

Furthermore, TiN coatings predominantly form a thin TiO_2 passive layer in NaCl solution, while TiBN coatings may additionally generate boron-containing oxides, contributing to enhanced barrier properties and improved resistance to chloride-induced degradation. In NaCl solution, TiCN coatings predominantly form a TiO_2 passive layer as the principal corrosion product, while localized degradation may occur at locations of coating defects, allowing chloride-induced corrosion processes at the coating/substrate interface (Figure 9). TiCN may exhibit a slightly higher defect density depending on deposition parameters, and carbon presence could influence the microstructure and electrical conductivity.

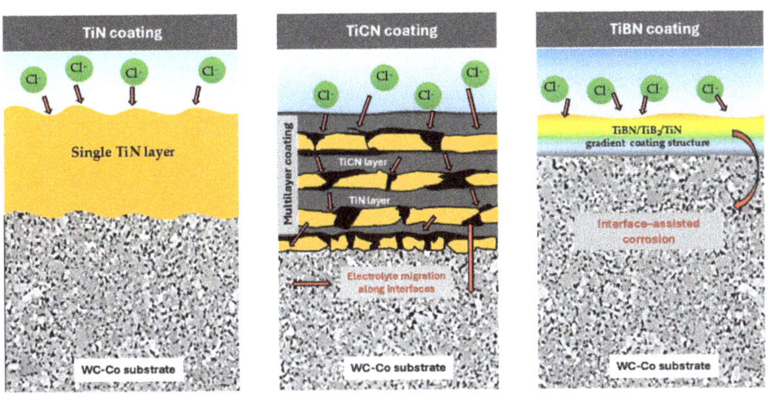

Figure 9. Schematic representation of corrosion mechanisms of different PACVD coatings on WC-10 wt.% Co substrate in 3.5 wt.% NaCl solution.

The corrosion performance of the TiN coating obtained in this study exhibited a very low corrosion current density, high polarization resistance and high charge transfer resistance, all indicating excellent barrier behaviour in NaCl solution. In a similar study,

https://doi.org/10.3390/coatings16030353

Chayuski et al. [43] reported significantly higher corrosion current densities in 3% NaCl for TiN coating deposited on a WC-Co substrate via arc-PVD with i_{corr} in the $\mu A/cm^2$ range. This suggests that the protective efficiency of the TiN coating strongly depends on the deposition method. In case of arc-PVD coating process, the presence of higher defect densities is often reported, as well as microdroplets occurrence. These defect locations could therefore act as a preferential location for electrolyte penetration and intensified corrosion. Matei et al. [44], who studied TiN, TiCN and TiAlN coatings on WC-Co substrates in identical electrolytes, also report the best corrosion resistance for TiN coating.

Nevertheless, the TiN coating in the present study demonstrates significantly lower corrosion current density (~0.011 $\mu A/cm^2$), placing it at the lower end of reported values for TiN on hardmetals in NaCl environments. The obtained polarization and charge transfer resistances are also significantly higher, confirming stronger barrier behaviour.

The superior performance of the TiN coating studied in this paper can therefore be attributed to the synergy of different factors, such as: moderate thickness (sufficient to act as good barrier), low surface roughness (which minimizes the potential of microscopic defect-induced corrosion), PACVD coating route (able to produce uniform, pore-less coating at low processing temperatures) and uniform single-layer architecture (which reduces the number of internal interfaces along which diffusion paths could form). Despite TiN coating showing lower adhesion quality compared to TiBN, its dense microstructure and lower permeability are crucial in governing corrosion response in NaCl environments.

5. Conclusions

While previous studies have extensively investigated the mechanical, tribological and structural properties of TiN, TiCN, and TiBN PACVD coatings on WC-Co substrates, a schematic comparison of their corrosion behaviour under the same conditions has not yet been reported. In particular, the influence of coating architecture and the deposition method on corrosion mechanisms and barrier performance in NaCl environments has not yet been reported. The corrosion behaviour of the investigated PACVD coatings in 3.5 wt.% NaCl solution is shown to be strongly influenced by coating architecture, structural integrity and composition. Distinct differences were observed for single-layer TiN coating and multilayer TiCN and TiBN coating:

1. The single-layer TiN coating exhibited the most noble corrosion potential, the highest polarization and charge transfer resistance and the lowest corrosion current density, all indicating the best corrosion resistance among the tested coating types. This can be attributed to the dense and chemically stable TiN layer that acts as a barrier for the electrolyte penetration and interfacial charge transfer, despite having lower hardness and adhesion in comparison to other coatings tested.
2. The TiBN gradient multilayer coating demonstrated intermediate corrosion resistance. Previous research showed this type of multilayer coating to have the highest hardness and excellent adhesion. Nevertheless, the increased number of internal interfaces arising from its specific architecture promotes moderate electrolyte penetration and interfacial charge transfer.
3. The multilayer TiCN coating showed the lowest corrosion resistance, characterized by the most negative corrosion potential, lowest polarization and charge transfer resistance and the highest corrosion current density. The interchanging TiN/TiCN layer architecture and higher carbon content lead to a less compact microstructure, causing increased interfacial capacitance. This creates conditions for electrolyte penetration and accelerated corrosion.
4. The obtained results show that the optimization of layer density and microstructural uniformity is essential for long-term corrosion resistance in the chloride-containing

https://doi.org/10.3390/coatings16030353

environments, despite complex coating architectures that exhibit superior mechanical and adhesion properties.

5. The PACVD TiN coating presented in this study has been shown to have superior corrosion resistance when compared to TiN coatings obtained by other deposition techniques reported in the literature, suggesting that using PACVD, a low-temperature coating method, significantly contributes to higher coating density and lower residual stresses and defects, which have been identified as crucial factors in the electrochemical resistance of the studied coatings.

Author Contributions: Conceptualization, M.Š. and M.K.; methodology, M.K., M.Š., M.S. and D.Ć.; formal analysis, M.Š., M.K. and D.Ć.; data curation, M.K., M.S. and M.Š.; writing—original draft preparation, M.Š., M.K. and D.Ć.; writing—review and editing, D.Ć., M.K. and M.Š. All authors have read and agreed to the published version of the manuscript.

Funding: This research received no external funding.

Institutional Review Board Statement: Not applicable.

Informed Consent Statement: Not applicable.

Data Availability Statement: The original contributions presented in this study are included in the article. Further inquiries can be directed to the corresponding author.

Conflicts of Interest: The authors declare no conflicts of interest.

Abbreviations

The following abbreviations are used in this manuscript:

PACVD	Plasma-Assisted Chemical Vapour Deposition
SCE	Saturated Calomel Electrode
EIS	Electrochemical Impedance Spectroscopy
LPR	Linear Polarization Resistance
HIP	Hot Isostatic Pressing
CPE	Constant Phase Element
AC	Alternating Current
DC	Direct Current
QDP	Quantitative Depth Profile
EEC	Equivalent Electrical Circuit

References

1. Acchar, W.; Gomes, U.U.; Kaysser, W.A.; Goring, J. Strength degradation of a tungsten carbide–cobalt composite at elevated temperatures. *Mater. Charact.* **1999**, *43*, 27–32. [CrossRef]
2. Wachowicz, J.; Bałaga, Z.; Podziewski, P. Wear characteristics of WC–Co cutting tools obtained by the U-FAST method during particleboard milling. *Materials* **2025**, *18*, 3907. [CrossRef]
3. Ren, X.; Zou, H.; Diao, Q.; Wang, C.; Wang, Y.; Li, H.; Sui, T.; Lin, B.; Yan, S. Surface modification technologies for enhancing the tribological properties of cemented carbides: A review. *Tribol. Int.* **2023**, *180*, 108257. [CrossRef]
4. Serra, M.; Batista, R.; Cinca, N.; Tarrés, E.; Jiménez-Piqué, E.; Moreno, P.; Llanes, L. Evaluation of fracture toughness of γ-phase-containing cemented carbides using through-thickness micronotches shaped by ultrashort pulsed laser ablation. *Crystals* **2025**, *15*, 279. [CrossRef]
5. Sheremet, I.; Akimov, G.; Andreiev, I.; Besarabets, Y.; Studenets, S.; Trosnikova, I.; Loboda, P. Phase and microstructure formation in WC–Co cemented carbide tools via cold isostatic pressing. *Materials* **2026**, *29*, 100865. [CrossRef]
6. Straumal, B.B.; Konyashin, I. Faceting/roughening of WC/binder interfaces in cemented carbides: A review. *Materials* **2023**, *16*, 3696. [CrossRef]
7. Fabijanić, T.A.; Kirin, T.; Šijaković, M.; Muminović, M.Š. Influence of Co content and chemical nature of the Co binder on the corrosion resistance of nanostructured WC–Co hardmetals in acidic solution. *Metals* **2021**, *11*, 1135. [CrossRef]

https://doi.org/10.3390/coatings16030353

8. Da Silva, F.S.; Benedetti, A.V.; Dosta, S.; Garcia-Cano, I.; Guilemany, J.M. Corrosion behavior of WC–Co coatings deposited by cold gas spray onto AA 7075-T6. *Corros. Sci.* **2018**, *136*, 231–243. [CrossRef]
9. Potgieter, J.H.; Whitefield, D.; Motsumi, V. Electrochemical corrosion behaviour of different grades of WC–Co, high-Cr white cast irons, and Hadfield steel in 1 M sulphuric acid. *Materials* **2021**, *14*, 6130. [CrossRef]
10. Human, A.M.; Exner, H.E. The relationship between electrochemical behaviour and in-service corrosion of WC-based cemented carbides. *Int. J. Refract. Met. Hard Mater.* **1997**, *15*, 65–71. [CrossRef]
11. Zheng, Y.; Huang, T.; Yao, J.; Fargas Ribas, G.; Armelín Diggroc, E.A.; Llanes Pitarch, L.M.; Xie, C.; Crespo Artiaga, D.; Bruna Escuer, P.; Lavigne, O. Electrochemical corrosion behavior of WC–Co and WC–Ni cemented carbides: Effect of chloride and sulfate ions. *Int. J. Refract. Met. Hard Mater.* **2025**, *132*, 107224. [CrossRef]
12. Schalk, N.; Tkadletz, M.; Mitterer, C. Hard coatings for cutting applications: Physical vs. chemical vapor deposition and future challenges for the coatings community. *Surf. Coat. Technol.* **2022**, *429*, 127949. [CrossRef]
13. Kazlauskas, D.; Jankauskas, V.; Tučkutė, S. Research on tribological characteristics of hard metal WC–Co tools with TiAlN and CrN PVD coatings for processing solid oak wood. *Coatings* **2020**, *10*, 632. [CrossRef]
14. Dabees, S.; Mirzaei, S.; Kaspar, P.; Holcman, V.; Sobola, D. Characterization and evaluation of engineered coating techniques for different cutting tools: A review. *Materials* **2022**, *15*, 5633. [CrossRef]
15. Ćorić, D.; Šnajdar Musa, M.; Sakoman, M.; Alar, Ž. Analysis of different complex multilayer PACVD coatings on nanostructured WC–Co cemented carbide. *Coatings* **2021**, *11*, 823. [CrossRef]
16. Sakoman, M.; Ćorić, D.; Šnajdar Musa, M. Plasma-assisted chemical vapor deposition of TiBN coatings on nanostructured cemented WC–Co. *Metals* **2020**, *10*, 1680. [CrossRef]
17. Šnajdar, M.; Ćorić, D.; Sakoman, M. Comparative study of multilayer hard coatings deposited on WC–Co hardmetals. *Coatings* **2024**, *14*, 674. [CrossRef]
18. Musa, M.Š.; Sakoman, M.; Ćorić, D.; Aleksandrov Fabijanić, T. Exploitation and wear properties of nanostructured WC–Co tool modified with plasma-assisted chemical vapor deposition TiBN coating. *Metals* **2021**, *11*, 333. [CrossRef]
19. Täschner, C.; Ljungberg, B.; Alfredsson, V.; Endler, I.; Leonhardt, A. Deposition of hard crystalline Al_2O_3 coatings by bipolar pulsed d.c. PACVD. *Surf. Coat. Technol.* **1998**, *108–109*, 257–264. [CrossRef]
20. Lamelas, V.; Bonvalet Rolland, M.; Walbrühl, M.; Borgenstam, A. Modelling the formation of detrimental phases in cemented carbides. *Mater. Des.* **2023**, *228*, 111823. [CrossRef]
21. Yuan, J.; Zhan, Q.; Huang, J.; Ding, S.; Li, H. Decarburization mechanisms of WC–Co during thermal spraying: Insights from controlled carbon loss and microstructure characterization. *Mater. Chem. Phys.* **2013**, *142*, 165–171. [CrossRef]
22. Formisano, A.; Capece Minutolo, F.; Caraviello, A.; Carrino, L.; Durante, M.; Langella, A. Influence of eta-phase on wear behaviour of WC–Co carbides. *Adv. Tribol.* **2016**, *2016*, 5063274. [CrossRef]
23. Movassagh-Alanagh, F.; Abdollah-Zadeh, A.; Alizadeh Zolbin, M.; Nemati, N.; Aghababaei, R. Plasma-enhanced chemical vapor deposition of TiB_2 and TiBN hard coatings using BBr_3. *Tribol. Int.* **2023**, *179*, 108137. [CrossRef]
24. Larsson, A.; Ruppi, S. Microstructure and properties of Ti(C,N) coatings produced by moderate temperature chemical vapour deposition. *Thin Solid Film.* **2002**, *402*, 203–210. [CrossRef]
25. Sakoman, M.; Ćorić, D.; Aleksandrov Fabijanić, T.; Kovačić, S. Tribological properties of coatings applied on near-nano and nanostructured WC–Co hardmetals by using plasma-assisted chemical vapour deposition technique. *Trans. FAMENA* **2020**, *44*, 29–42. [CrossRef]
26. Su, Y.L.; Kao, W.H. Tribological behavior and wear mechanisms of TiN/TiCN/TiN multilayer coatings. *J. Mater. Eng. Perform.* **1998**, *7*, 461–468. [CrossRef]
27. Laugier, M.T. Adhesion of TiC and TiN coatings prepared by chemical vapour deposition on WC–Co-based cemented carbides. *J. Mater. Sci.* **1986**, *21*, 1507–1514. [CrossRef]
28. Ruan, H.; Wang, Z.; Wang, L.; Sun, L.; Peng, H.; Ke, P.; Wang, A. Designed Ti/TiN sub-layers suppressing the crack and erosion of TiAlN coatings. *Surf. Coat. Technol.* **2022**, *438*, 128419. [CrossRef]
29. Ali, R.; Sebastiani, M.; Bemporad, E. Influence of Ti–TiN multilayer PVD-coatings design on residual stresses and adhesion. *Surf. Coat. Technol.* **2015**, *276*, 380–389. [CrossRef]
30. García, J.A.; Claver, A.; Marques, M.; Almandoz, E.; Fernández de Ara, J.; Palacio, J.F.; Azkona, I. Improvement of the tribocorrosion properties of cemented carbide (WC–TiC–Co) samples with PVD coating. *Coatings* **2022**, *12*, 1884. [CrossRef]
31. Wagner, A.; Übleis, C.; Krawinkler, M.; Dipolt, C.; Strobl, V. Corrosion resistance and tribological behavior of PACVD coated and gas nitrided surfaces as alternative to hard chromium for piston rod coatings. *BHM Berg- Hüttenmänn. Monatsh.* **2021**, *166*, 458–471. [CrossRef]
32. Han, B.; Dong, W.; Fan, B.; Zhu, S. Comparison on the immersion corrosion and electrochemical corrosion resistance of WC–Al_2O_3 composites and WC–Co cemented carbide in NaCl solution. *RSC Adv.* **2021**, *11*, 22495–22507. [CrossRef]
33. Ismail, A.; Aziz, N. Corrosion behavior of WC–Co and WC–Ni in 3.5% NaCl at increasing temperature. *Appl. Mech. Mater.* **2014**, *660*, 135–139. [CrossRef]

https://doi.org/10.3390/coatings16030353

34. *EN 1071-2:2002*; Advanced Technical Ceramics—Methods of Test for Ceramic Coatings—Part 2: Determination of Coating Thickness by the Crater Grinding Method. iTeh: Newark, DE, USA, 2002.
35. Belhamra, N.; Chala, A. Corrosion behavior of TiN-coated cutting carbide tools. *Main Group Chem.* **2014**, *13*, 65–71. [CrossRef]
36. López-Ortega, A.; Arana, J.L.; Bayón, R. Tribocorrosion of passive materials: A review on test procedures and standards. *Int. J. Corros.* **2018**, *2018*, 7345346. [CrossRef]
37. Yuan, X.; Yu, R.; Du, Y.; Lei, B. EIS study on the barrier properties of silicone–epoxy coatings enhanced by different PDA-BN@ZnO nano-pigment contents. *Corros. Sci.* **2024**, *228*, 111810. [CrossRef]
38. Bozzini, B.; Tavola, F.; Travella, A.; Alleva, A.; Mele, C.; Emanuele, E.; Tedeschi, S.; De Gaudenzi, G.P. In-depth understanding of hardmetal corrosion performance reveals a path to the electrochemical demolition of scrap. *Metals* **2023**, *13*, 1376. [CrossRef]
39. Balbino, N.A.N.; Corrêa, E.O.; Huanca, D.R.; de Freitas Matos, F.A.; de Carvalho Valeriano, L. Comparative study of corrosion behaviors of WC–NiMo and WC–Co cemented carbides. *Materials* **2023**, *16*, 4480. [CrossRef] [PubMed]
40. *EN ISO 17475:2019*; Corrosion of Metals and Alloys—Electrochemical Test Methods—Guidelines for Conducting Potentiostatic and Potentiodynamic Polarization Measurements. ISO: Geneva, Switzerland, 2019.
41. Aleksandrov Fabijanić, T.; Šnajdar, M.; Kurtela, M.; Šimunović, V.; Marciuš, M.; Klaić, M. Corrosion Resistance of Nanostructured Cemented Carbides with Alternative FeNi and FeNiCo Binders. *Nanomaterials* **2023**, *13*, 1407. [CrossRef]
42. Macdonald, J.R. (Ed.) *Impedance Spectroscopy*; John Wiley & Sons: New York, NY, USA, 1987.
43. Chayeuski, V.; Zhylinski, V.; Kazachenko, V.; Tarasevich, A.; Taleb, A. Structural and Mechanical Properties of DLC/TiN Coatings on Carbide for Wood-Cutting Applications. *Coatings* **2023**, *13*, 1192. [CrossRef]
44. Matei, A.; Pencea, I.; Branzei, M.; Trancă, D.E.; Ţepeş, G.; Sfăt, C.E.; Ciovica (Coman), E.; Gherghilescu, A.I.; Stanciu, G.A. Corrosion resistance appraisal of TiN, TiCN and TiAlN coatings deposited by CAE-PVD method on WC–Co cutting tools exposed to artificial sea water. *Appl. Surf. Sci.* **2015**, *358*, 572–578. [CrossRef]

Disclaimer/Publisher's Note: The statements, opinions and data contained in all publications are solely those of the individual author(s) and contributor(s) and not of MDPI and/or the editor(s). MDPI and/or the editor(s) disclaim responsibility for any injury to people or property resulting from any ideas, methods, instructions or products referred to in the content.

https://doi.org/10.3390/coatings16030353

MDPI AG
Grosspeteranlage 5
4052 Basel
Switzerland
Tel.: +41 61 683 77 34

Coatings Editorial Office
E-mail: coatings@mdpi.com
www.mdpi.com/journal/coatings

Disclaimer/Publisher's Note: The title and front matter of this reprint are at the discretion of the Guest Editors. The publisher is not responsible for their content or any associated concerns. The statements, opinions and data contained in all individual articles are solely those of the individual Editors and contributors and not of MDPI. MDPI disclaims responsibility for any injury to people or property resulting from any ideas, methods, instructions or products referred to in the content.

www.ingramcontent.com/pod-product-compliance
Lightning Source LLC
Chambersburg PA
CBHW062042220926
51850CB00010B/2640
9783725875955